Urban Neighbourhood Formations

This book examines the formation of urban neighbourhoods in the Middle East, Africa, and South Asia. It departs from 'neighbourhoods' to consider identity, coexistence, solidarity, and violence in relations to a place.

Urban Neighbourhood Formations revolves around three major aspects of the making and unmaking of neighbourhoods: spatial and temporal boundaries of neighbourhoods, neighbourhoods as imagined and narrated entities, and neighbourhood as social relations. With extensive case studies from Johannesburg to Istanbul and from Jerusalem to Delhi, this volume shows how spatial amenities, immaterial processes of narrating and dreaming, and the lasting effect of intimacies and violence in a neighbourhood are intertwined and negotiated over time in the construction of moral orders, urban practices, and political identities at large.

This book offers insights into neighbourhood formations in an age of constant mobility and helps us understand the grassroots-level dynamics of xenophobia and hostility, as much as welcoming and openness. It would be of interest for both academics and more general audiences, as well as for students of undergraduate and postgraduate courses in urban studies and anthropology.

Hilal Alkan is a research fellow of the Alexander von Humboldt Foundation at Leibniz-Zentrum Moderner Orient, Berlin. Her work focuses on charitable giving, migration, gender, and social welfare, through the lenses of anthropology, citizenship studies, and urban studies. Her publications include 'The Gift of Hospitality and (Un)welcoming Syrian Migrants in Turkey', *American Ethnologist* (2019) and 'The Sexual Politics of War: Reading the Kurdish Conflict through Images of Women', *Les Cahiers du CEDREF* (2018).

Nazan Maksudyan is Professor of History at the Freie Universität Berlin and a research associate at the Centre Marc Bloch, Berlin. Her research focuses on the history of children and youth, with special interest in gender, sexuality, education, humanitarianism, and non-Muslims. Her publications include *Women and the City, Women in the City: A Gendered Perspective on Ottoman Urban History*, ed. (2014); *Ottoman Children & Youth During WWI* (2019), 'Orphans, Cities, and the State: Vocational Orphanages (*Islahhanes*) and "Reform" in the Late Ottoman Urban Space', *IJMES* 43 (2011).

Routledge Studies in Urbanism and the City

For more information about this series, please visit www.routledge.com/Routledge-Studies-in-Urbanism-and-the-City/book-series/RSUC

Urban Neighbourhood Formations

Boundaries, Narrations and
Intimacies

**Edited by Hilal Alkan and
Nazan Maksudyan**

LONDON AND NEW YORK

First published 2020
by Routledge
2 Park Square, Milton Park, Abingdon, Oxon OX14 4RN

and by Routledge
605 Third Avenue, New York, NY 10017

Routledge is an imprint of the Taylor & Francis Group, an informa business

First issued in paperback 2021

Publisher's Note
The publisher has gone to great lengths to ensure the quality of this reprint but points out that some imperfections in the original copies may be apparent.

British Library Cataloguing-in-Publication Data
A catalogue record for this book is available from the British Library

Library of Congress Cataloging-in-Publication Data
A catalog record has been requested for this book

ISBN 13: 978-0-367-25510-7 (hbk)
ISBN 13: 978-1-03-223805-0 (pbk)
ISBN 13: 978-0-429-28814-2 (ebk)

Typeset in Times New Roman
by Swales & Willis, Exeter, Devon, UK

Contents

PART III
Intimacies: neighbourhoods as sources and objects of claim-making   159

Figures

Editor and contributor biographies

Hilal Alkan is a Georg Forster Fellow of the Alexander von Humboldt Foundation at the Leibniz-Zentrum Moderner Orient. She received her PhD in Political Science from The Open University, UK. She was a EUME Fellow of Forum Transregionale Studien in Berlin in 2016–2017 and the receiver of Potsdam University's Voltaire Award for Tolerance, International Understanding and Respect for Tolerances in 2017. Her work focuses on charitable giving, migration, and social welfare, through the ' lenses of anthropology, citizenship studies, and urban studies. In her recent project called 'The Dyad of Care and Discipline: Aiding Syrian Migrants in Turkey and Germany', she is working on informal neighbourhood initiatives that aid Syrian migrants in their settlement in both contexts and comparing migrants' experiences in both countries. Her wider research interests include gendered experiences of war and violence, feminist ethics of care, and plant–human relations.

Nazan Maksudyan is Einstein Guest Professor at the Freie Universität Berlin and a research associate at the Centre Marc Bloch, Berlin. She was a EUME Fellow 2009–2010 at the Wissenschaftskolleg zu Berlin and an Alexander von Humboldt Stiftung Postdoctoral Fellow at the ZMO in 2010–2011. Her research mainly focuses on the history of children and youth in the late Ottoman Empire, with special interest in gender, sexuality, education, humanitarianism, and non-Muslims. Among her publications are *Orphans and Destitute Children in the Late Ottoman Empire* (Syracuse University Press, 2014), ed. *Women and the City, Women in the City* (New York: Berghahn, 2014); *Ottoman Children and Youth During World War I* (Syracuse University Press, 2019), 'Orphans, Cities, and the State: Vocational Orphanages (Islahhanes) and "Reform" in the Late Ottoman Urban Space', *IJMES* 43 (2011); 'Foster-Daughter or Servant, Charity or Abuse: Beslemes in the Late Ottoman Empire', *Journal of Historical Sociology* 21 (2008).

***

Johannes Becker is a Research Fellow at the Center of Methods in Social Sciences, University of Goettingen, and coordinator of the research project 'Dynamic Figurations of Refugees, Migrants, and Long-time Residents in Jordan since 1946', sponsored by the German Research Foundation.

Aatina Nasir Malik is a PhD candidate in the Department of Humanities and Social Sciences at the Indian Institute of Technology, Delhi. Her research interests include political violence, Kashmir studies, gender, and the everyday. Her PhD thesis explores political violence and the everyday in Kashmir, underscoring the way political violence makes and unmakes social life in a neighbourhood in Srinagar, and thereon how different subjectivities emerge.

Ahmad Moradi received his PhD in Social Anthropology at the University of Manchester, UK. His thesis titled 'Politics of Persuasion: Making and Unmaking Revolution in Iran', explores the intersection of sacrifice and citizenship in the Islamic Republic of Iran. His research interests include collective and individual experience of loss, welfare and networks of care, political anthropology, and the anthropology of the Middle East. He is also involved in the practice of creative ethnography, the result of which has been an artistic monograph on the subject of illegal migration and exhibitions in Tehran, Budapest, and Amsterdam.

Derya Özkaya is a doctoral researcher at the Otto Suhr Institute of Political Science at Freie Universität Berlin. She is currently working on the role of emotional and affective dynamics for political participation and transformation exploring the possibilities and limitations of change starting from the Gezi uprisings of 2013 in Turkey. Her research interests include social movements, collective memory, commemorative practices, revolutionary movements in Turkey, collective emotions, and political affect.

Samprati Pani is a PhD candidate at the Department of Sociology, Shiv Nadar University, Uttar Pradesh. Through intersections of urban informality, design, and spatial practices, her research focuses on the making and techniques of weekly bazaars in Delhi. She has a Master's and MPhil in Sociology from Delhi School of Economics, University of Delhi. She has worked in academic publishing as an editor for many years and is also a trained graphic designer. Her interests include ordinary aesthetics, urban spatialities, street typography, and practices of street vending. She blogs on various aspects of the everyday city at chiraghdilli.wordpress.com.

Kholofelo Rameetse holds a BSc Honours degree in Geography and Environmental Studies from the University of the Witwatersrand, Johannesburg. Over the past three years, she has conducted her MSc research in Eldorado Park and neighbouring locations. Her research interests are post-apartheid urban development policies and practices. She is also a business development associate for an independent renewable energy developer.

Annegret Roelcke is a research fellow at Leibniz-Zentrum Moderner Orient Berlin and a PhD candidate at the Institut für Islamwissenschaft at Freie Universität Berlin. Her research investigates identity politics in urban settings in Turkey and the Middle East, combining anthropological and historical approaches. Currently she is working on her dissertation on contemporary urban heritage politics in Istanbul. She studied Arab and Islamic Studies at Leipzig University, Damascus University, Boğaziçi University (Istanbul), and Freie Universität Berlin, from where she holds an MA in Islamic Studies. In 2016, she received a residency scholarship by the German Orient Institut Istanbul.

Samuli Schielke is a social and cultural anthropologist working on contemporary Egypt. He is a research fellow at Leibniz-Zentrum Moderner Orient (ZMO) and an associate primary investigator at Berlin Graduate School of Muslim Cultures and Societies. He is the author or editor of *The Perils of Joy* (2012), *The Global Horizon* (with Knut Graw, 2012), *Ordinary Lives and Grand Schemes* (with Liza Debevec, 2012), *Egypt in the Future Tense* (2015), and *Migrant Dreams* (2020).

Alex Wafer is Senior Lecturer in the School of Geography, Archaeology and Environmental Studies at the University of the Witwatersrand, Johannesburg. He is currently a co-investigator in the Scottish Funding Council/ODA Global Challenges Internal Fund 2019 'Platforms for Precarity: Assessing the Development of Digital Platforms to Support Unemployed, Unbanked or Low Income Precarious Workers in the Informal Economy in South Africa', researching the impacts of digital technologies on the geographies of work and livelihoods in South Africa.

Yair Wallach is a senior lecturer in Israeli Studies at SOAS, University of London. His research deals with social and cultural history of modern Palestine/Israel, focusing on visual and material culture and the urban sphere. He has written on the use of maps as national icons, on the sociology of late Ottoman Palestine's Jewish communities, and on 'shared space' in Jerusalem before 1948. His book *A City in Fragments: Urban Text in Modern Jerusalem* (Stanford University Press, forthcoming 2020) looks at Arabic and Hebrew street texts in the turbulent context of Jerusalem between 1850 and 1948.

Urszula Ewa Woźniak is a doctoral candidate in sociology and a research associate at the Department for Social Sciences at Humboldt-Universität zu Berlin. Her research interests include political sociology and political anthropology, migration and transnationalism, urban studies, and Turkish studies. With her ethnographic research on two Istanbul neighbourhoods, she examines the relevance of the neighbourhood as a contested system of order both in the realm of everyday life and political discourse. Her PhD project traces how, in the context of resurgent nationalism under the current neoliberal Islamic rule, the increasing polarisation of Turkish society reverberates in the making of the neighbourhood as moral territory.

Sezai Ozan Zeybek is a Freudenberg Stiftung Fellow at the Forum Transregionale Studien in Berlin. He received his PhD in Geography from The Open University in the UK. He was formerly an associate professor in the Sociology Department at Istanbul Bilgi University. He works on more-than-human collectives and ecologies, state violence, and militarism. His recent publications include a study of stray dogs in Istanbul, livestock killings in the Kurdish region of Turkey during the forced displacement of the Kurdish population in the 1990s, and the criminalisation of goats during the appropriation of forests in the early Republican period of Turkey. He has also published three children's stories.

Acknowledgements

We would first like to thank the Alexander von Humboldt Foundation Postdoctoral Research fellowship programmes, which allowed both of us to solely focus on our research and writing. We are also grateful to our host institute at the time, Leibniz-Zentrum Moderner Orient (ZMO), which provided the necessary support and encouragement for us to focus on the topic of neighbourhoods.

This edited volume is the product of the workshop 'Neighbourhoods in Times of Change and Crisis', organised by Hilal Alkan and Nazan Maksudyan in Berlin in May 2018, as part of the yearly activities of the research group at the Leibniz-Zentrum Moderner Orient, 'Cities as Laboratories of Change'. We are indebted to our exquisite group of colleagues working on different aspects of urban studies. Several members of this research group, including Ulrike Freitag, Nora Lafi, Sanaa Alimia, İlkay Yılmaz, Deniz Yonucu, Claudia Ghrawi, and Annegret Roelcke, contributed greatly to our discussions while developing the idea of a meeting around neighbourhoods.

We also benefited greatly from the event 'Changing Neighbourhoods', that we curated together with Nadia von Maltzahn and Birgit Schäbler (both from Orient Institut Beirut) with the cooperation of the Max Weber Stiftung – Deutsche Geisteswissenschaftliche Institute im Ausland, the Forum Transregionale Studien, the Orient Institut Beirut, and the Leibniz-Zentrum Moderner Orient. In a one-day open discussion session, we have dealt with a number of issues including the social production of neighbourhoods, self-organisation and parallel infrastructures, conflictual moments, and gendered use of spaces.

We would also like to thank all the contributors of this volume who kindly revised their chapters based on our suggestions and comments and diligently worked with the volume's theoretical frameworks so as to create a very solid contribution to the available literature. We are also grateful to the editorial team of the Routledge Studies in Urbanism and the City for their hard work and punctuality.

Finally, our special thanks go to publications' coordinator of the ZMO, Svenja Becherer for her support, our copy-editor Andy Tarrant for his diligence, and student assistant Diana Gluck for her attentive eye and being available at the last minute.

Introduction

Hilal Alkan and Nazan Maksudyan

This volume explores the themes of 'neighbours' and 'neighbourhoods' as a point of departure for considering identity, coexistence, solidarity, and violence and their relationship to place in the Middle East, Africa, and South Asia. Neighbourhood as an intimate place with blurred yet shared boundaries in the mental-maps of its residents is home to conflicts over space and time. Its histories, material formations, as well as transformations are both indicative and inductive of micro and macro social phenomena. Historically as much as recently, neighbourhoods have been the loci of resistance and political dissidence, but also coexistence of potentially hostile groups. Therefore, envisioning the neighbourhood as more than an administrative category aids in understanding spatial formations of the social and the political in modern urban settings.

This collection brings together anthropological, historical, and urban studies perspectives with articles focusing on neighbourhoods in Istanbul, Alexandria, Johannesburg, Jerusalem, Delhi, Srinagar, Bandar Abbas, and the small town of Malkara in Western Turkey. The chapters approach neighbourhoods as places in the making, with references to the above-mentioned tensions and contestations. Based on empirical research (ethnography, life story interviews, and archival research), they serve to envisage a concept of the neighbourhood that is neither static nor ephemeral. In that sense, they contribute to the urban studies literature by re-introducing a concept that is very much alive in contemporary cities and the lives of urban dwellers.

While working on this collection, we, the editors and contributors alike, have many times stumbled on the same question, 'What actually is a neighbourhood?', although we have all worked extensively in and on neighbourhoods for many years. The collective endeavour of this project can, therefore, also be read as a search for answers (not a singular answer) to this question—answers that introduce a variety of angles to both posing the question and replying to it.

The first set of answers introduce a spatial angle and discuss the *boundaries* that delimit neighbourhoods. Here the function of boundaries is less to separate and more to make a certain neighbourhood identifiable from the rest of the city. The process of identification may refer to residence histories, everyday practices, conceptual registers, or to embodied ethnic differences. The boundaries, therefore, do not have to be material (walls, railways, highways) but in any case, they are materialised in the bodies, in the language, and in the daily practices of neighbourhood dwellers and outsiders to make each and every neighbourhood knowable, recognisable, and identifiable.

The second set of answers introduce a *narrative* angle. Neighbourhoods are not only lived in and experienced, they are also imagined and narrated. Stories situate neighbourhoods between a past and a future, temporally rooting them in the city. They also attribute textures and flavours to them. Yet, stories also fiercely compete with each other. Some become dominant—even official—in the end, while others are completely forgotten; just like the neighbourhoods they talk about.

Finally, the last set gives us the *social* relations angle. Whatever the legal title or administrative boundaries, it is the neighbours, as social agents, who turn a piece of urban settlement into a neighbourhood. They make and remake the place by their interactions with each other. These interactions involve care and solidarity as much as conflict and hostility. They also involve making clashing claims over the ownership and moral order of the neighbourhood.

Through these angles this collection explores *urban neighbourhood formations* in the Middle East, Africa, and South Asia and illustrates how spatial boundaries, immaterial processes of narrating and dreaming, and the lasting impact of affective ties in a neighbourhood are intertwined and negotiated over time in the construction of moral orders, urban practices, and political identities at large. Through this diversity we hope to achieve a fuller picture of 'what a neighbourhood actually is'—a richly coloured picture, vivacious and full of nuances.

Comparative urbanism on a micro scale

The chapters in this collection contribute to the discussion that has been taking place around the concept of 'comparative urbanism' during the past decade (see for example the special issue of *Urban Geography* 2016, 33(6) and the debate in the *International Journal of Urban and Regional Research* 2016, 40(1)). By approaching neighbourhoods in their translocal interconnectedness, the collection in its entirety carries comparative urbanism to a micro scale. It adds insights to this growing literature through chapters that are aware of their locatedness, but which still remain open to comparison, along with offering global insights. In that respect, we follow Robinson's approach 'to start conceptualization from any city and to draw insights from a wide array of contexts while acknowledging the locatedness of all theoretical endeavour' (Robinson 2016, 188). We also stress the need to reconfigure the conventions surrounding the determination of the 'third term' (Jacobs 2012) that makes a comparative approach across numerous urban contexts, and across geographic and temporal distances, possible. The chapters, focusing on urban neighbourhoods as locally configured, yet interconnected spaces, both play with and go beyond the centre–periphery, global–local, and structure–agency binaries.

Urban studies scholars have long had lengthy discussion on what constitutes a city. What are the necessary 'ingredients of a city', and of 'cityness'? In a recent contribution in their 'search for urbanity', Alimia et al. (2018) point at 'urban space and its infrastructure' (the walls, neighbourhoods, roads, flyovers, and street lighting), 'the residents of the urban space' (their behaviours, daily rhythms, emotions, and sensory experiences), and the 'governance techniques' (from above) as indicators of urbanity. However, we believe that given the size, expanse, and heterogeneity of twenty-first century cities, a monolithic yet sound analysis of an entire city is not quite viable. Stressing the need to re-focus urban studies on a micro scale that does not prejudge the emancipatory nature of local identities, we advocate a reduction in scale to maintain the complexity of micro aspects and the agency of the actors. The micro focus on neighbourhoods is targeted at demonstrating the local processes involved in identity formation, elucidating forms of social cohesion (or lack thereof), and analysing locally situated conflicts vs. peaceful coexistence.

In light of the new discussions around the 'spatial turn' (Döring and Thielmann 2015; Warf and Arias 2008) and micro-spatial methodologies (De Vito and Gerritsen 2017; Epple 2012), focusing on the 'micro scale' better addresses social and cultural aspects, the agency of the actors involved, and the differentiations around class, ethnicity, religion, and gender. Micro-spatiality relies on an epistemology of difference and connectedness at the same time. In other words, it offers a bottom-up analysis of the comprehension of the global, to illustrate the dialogue between micro and transnational processes.

This edited volume also further qualifies these discussions by approaching cities not as wholes—even as heterogeneous as they can be—but as fragmented and fractured spatial entities. We posit that neighbourhood as a unit and as a concept imminently reproduces heterogeneity and challenges a single and unified identity. By focusing on minor components of cities—i.e. neighbourhoods—we aim to take a closer look at the dynamics of memory, hostility, competition, and conviviality that play a major role in the make-up of cities. This close-up view, while making possible a nuanced understanding of the daily affairs that constitute urbanism as a way of life (Wirth 1938), also challenges the conceptions of the urban as the locus of a blasé attitude and distantiated relations (Simmel [1905] 2012). Yet the chapters also refuse to give in to a romanticised view of the neighbourhoods in which they are seen as the locations of enclosed and close-knit communities by showing how extra-neighbourhood actors (such as planning agencies, municipalities, armed groups, and global financial capital) have huge impacts on neighbourhoods in negotiation with and sometimes to the dismay of their residents.

As Doreen Massey noted, space is 'constructed out of the multiplicity of social relations across all spatial scales' (1994, 4). From a dynamic comparative approach, the contributions in this volume approach the neighbourhood as an open space, the specificity of which stems from distinctive processes of place-making via the connections and conflicts that operate within and beyond each place, rather than through bounded and self-referential identities. However, the neighbourhood is also of a very particular scale, 'the right scale' as Veena Das puts it (2017, 194), which makes it possible to see the entanglements of these various multi-scalar vectors in their most concrete form. Neighbourhood, unlike many other social science concepts, is not abstract, but a 'real collective level of lives as lived' (Rechtman 2017). As such it provides the ideal scale and setting for the questions this collection poses.

Narrowing down the scale also gives us the advantage of methodological heterogeneity. While the majority of our contributors employed ethnographic methods in their endeavours to disentangle the complexities that neighbourhoods bear, the micro scale of the neighbourhood proved equally suitable to archival research, and narrative and discourse analysis. Through these methods, they managed to explore not only existing memories, stories, and neighbourhoods themselves, but also the lack of them: the silence and hollowed-out spaces. Employing a micro-spatial approach also gave them the advantage of expanding the chronological scope and connecting current neighbourhood formations to historical formations of equal or larger scale. The transdisciplinary research methodologies employed by our contributors made it possible for them to reflect on questions such as community formation, social cohesion, and violence in a locality, in their relation to colonial history, nation-building, and biopolitics.

These questions and their connections to wider social and historical phenomena are highly pertinent, not only to the methodologies but also to the cases chosen for this book. The cities in which the neighbourhoods are located have colonial history materially written on them (Alexandria, Delhi, Johannesburg, Srinagar, Jerusalem). They have long been the object of orientalist fantasies (Istanbul, Delhi, Alexandria, Jerusalem) and carry the tensions of nation-building with a new moral order (Bandar Abbas, Istanbul, Malkara, Srinagar). Yet all of them are coeval, not only as beings of the present but also as formations of the present marked by raging neoliberalism, the threat of climate change, and increasing movement (of humans, materials, money, and information).

The selection of these cities reflects two concerns. First, the need for contributions coming from the Global South to revise basic premises of urban theory and, second, the necessity of making South-to-South comparisons and translations. Inspired by bottom-up perspectives on the comprehension of the global, we stress the importance of both transnational processes and the histories of diverse places in their interconnectedness.

All of the neighbourhoods in this volume are located in the cities of the Global South, and most in 'megacities'. In the literature, the image of the megacity—the chaotic, ungraspable and ever-growing conglomerate, characteristic of unregulated and unplanned Third World development—sets the limits of ethnographic representation and scholarly comprehension. This megacity often has a metonym: the slum (Roy 2011). However, the neighbourhoods covered in this collection resist this image of extraordinariness and being beyond representation: they are 'ordinary' neighbourhoods, not necessarily marked by poverty, extreme violence, or revolt. They are not the 'typical slums' with muddy, narrow alleys and make-shift houses. They are all lower to middle class (or even upper class) neighbourhoods with well-entrenched positions in their cities. They are not places where a Southern exceptionalism could be sought. They are illustrative of tensions that are unique in one sense—that they are spatially and temporally specific—and equally global and 'ordinary' (Robinson 2006) in another sense.

Neighbourhood formations

This collection is organised into three parts that respectively refer to three major elements of neighbourhood formations: borders, stories, and intimacies. These have also been significant themes in the domain of cultural production that concern neighbourhoods in the recent years. We introduce each part with reference to such an event, which we came across during the production of this book—from its emergence as an idea till its completion. The three parts of the book also bring together three fields within urban studies that are generally regarded as separate—namely, infrastructure and public space; memory studies; and social relations—and discuss them in relation to each other as dimensions of making and unmaking neighbourhoods. As a result of this sophisticated interrelatedness many chapters within the volume could potentially be placed under more than one subsection. Engaged readers will also not fail to find more cross-cutting themes than we point out in this introduction.

Borders: material, temporal and conceptual boundaries of neighbourhoods

A line from Robert Frost's poem, 'Mending Wall' was made famous when Ai Weiwei named his 2017 New York City-wide public art project, through which he commented on rising borders in the face of the increased refugee influx to the West, 'Good fences make good neighbours'. In the poem, the narrator and his neighbour work on rebuilding the damaged wall that marks the boundary between their fields. The narrator thinks they do not really need a wall, as it is obvious where one's field ends and the other's begins. The neighbour, however, constantly repeats the phrase, 'Good fences make good neighbours'. For the narrator, the phrase is not self-explanatory; for the neighbour, it is—referring to 'establishing boundaries' between neighbours so that both know where to stop and respect each other's space. Yet there is hostility in his voice and the way he grabs the stones: to raise the wall or to throw them at the narrator? We do not know, but we feel the threat.

The notion of neighbourhood immediately connotes proximity; but it also invites the question of how to manage that proximity, how to establish boundaries. The first part of our volume approaches the formation of neighbourhoods from this angle of boundaries, and the conflicts and violence immanent within them. These borders/boundaries can be material or invisible, temporal or conceptual, gendered or ethnic; they can be established between neighbours or between neighbourhoods.

Borders, first of all, refer to the question of identifiability. Approaching the city as composed of geographically identifiable neighbourhoods with clearly demarcated borders is a very particular notion that was created at the turn of the twentieth century by American urban planners such as William E. Drummond. Neighbourhoods as standardised geographical units were imagined as the subsections of the greater whole, the city. What we and our contributors see when looking at neighbourhoods, on the other hand, is often a lack of geographical precision, let alone the replicability of neighbourhood characteristics. In that sense, the contributions in this part (and throughout the volume) follow Robert J. Chaskin, who commented that,

> There is, however, power in the idea of the neighbourhood, power that comes not from its precision as a sociological construction but from its nuanced complexity as a vernacular term. Neighbourhood is known, if not understood, and in any given case, there is likely to be wide agreement on its existence, if not its parameters.
>
> (1997, 523)

Our contributors approach the paradoxically simultaneous existence of an awareness of neighbourhood boundaries and the lack of geographical precision (being known but not understood) as a matter of negotiation between residents and state authorities, and between neighbourhoods.

Alex Wafer and Kholofelo Rameetse's contribution to this volume, 'What makes a township a neighbourhood? The case of Eldorado Park, Johannesburg' illustrates how such tensions are aggravated by new real estate developments and infrastructural investments on the border between two neighbourhoods, both of which carry the legacy of apartheid in their racial configuration. Within the post-apartheid state of abandonment, the designated 'brown' township of Eldorado Park is in fierce competition with the neighbouring 'black' township of Kliptown over the crumbs of public resources directed to townships. The tension finds its expression in resentment of the new racial hierarchies of South Africa, as much as it refers to a politico-spatial configuration that positions one neighbourhood against the other.

In Sezai Ozan Zeybek's chapter, 'Killing time in a Roma neighbourhood: habitus and precariousness in a small town in Western Turkey', we see how the neighbourhood identifiers, i.e. the borders, are carried around on the bodies of its residents. As soon as the Roma youth of the neighbourhood cross the threshold to the other (Turkish) neighbourhood, they are quickly chased off or met with racist hostility. The unwelcoming attitude is so stark that the parameters could not be any more precise if there were a wall between the two neighbourhoods. Zeybek's chapter offers an analysis of this wall, built along the lines of class and ethnicity. Zeybek also analyses temporality as another layer of this wall and looks into the temporal formations of the neighbourhood. Living there, according to many of its inhabitants, was ordinary, mundane, and repetitive; this was a place where 'nothing happens', as opposed to other places that are marked by dynamism, vitality, and change. Doing this, the Roma residents of the neighbourhood refer to certain spatial hierarchies that attribute superfluity to the peripheries, to the provincial places. This experience of 'dead time', Zeybek argues, impacts the texture of the neighbourhood and marks its co-constitution within globally entangled spatial and temporal configurations.

The boundaries that define and delimit neighbourhoods are often tangible, material entities that are built upon or translate into infrastructures, spatial knots, and amenities. However, as we see in Samprati Pani's chapter, 'Of Basti and bazaar: place-making and women's lives in Nizamuddin, Delhi', these boundaries are also made through daily and persistent practices. Pani illustrates that a spatial heart is key to the formation of the Delhi neighbourhood of Nizamuddin: the weekly market, *the bazaar* as Pani's interlocutors call it; and that this formation is only realised via the participation and activities of neighbours. Pani deals with the entangled making of these two places, the weekly market and the neighbourhood. The bazaar for her is not only the setting of intense social relations, but is also constitutive of the social networks through which we can make sense of the micro-politics of community formation. Neighbourhood is made through relationships, as much as the relationships are built within and through places. The chapter offers ground from which to view the multiple, complex, and fluctuating relationships that exist between the individuals, spaces, and structures involved in the social and spatial construction of place. Pani uses place-making as an analytical lens to explore lived experiences of neighbours and to connect social relations with spatial formations, and thus, emphasises the creative elements of human action and interaction (Friedmann 2010).

In Pani's ethnographic research, the place-makers are the women of the neighbourhood. They turn the inner-city, lower-class neighbourhood of Nizamuddin into a familiar and hospitable home for their dense social network. Their place-making practices are woven around one key activity (among possible others): visiting the neighbourhood's weekly market. Pani's emphasis on women's routines and practices and Zeybek's sole focus on the young Roma men are not random choices. Both authors are deliberate in their attempts to show how gender is an important parameter in neighbourhood formations. Neighbourhoods are inescapably gendered in many senses: in their spatial and temporal configuration, in terms of the relations they give way to and finally in the moral conflicts they accommodate. People with different gender identities and sexual orientations both experience and shape neighbourhoods in radically different ways. Although gender is expressly present in these two chapters, it is also a cross-cutting theme that finds resonance in Johannes Becker's contribution on Jerusalem, Aatina Nasir Malik's work on Srinagar, and Urszula Ewa Woźniak and Hilal Alkan's discussions on different Istanbul neighbourhoods.

One final boundary, the limits of the concept itself, is discussed by Alex Wafer and Kholofelo Rameetse. They illustrate at length 'the out-of-placeness' of the term neighbourhood in their case study, the 'township' of Eldorado Park. The authors argue that neighbourhoods do not really exist in South African cities. There are, instead, townships and suburbs. Wafer and Rameetse examine why. However, despite the many things that make South African cities peculiar, they do not intend to follow the 'exceptionalism' school that dominates urban studies on South African cities. Thinking comparatively across contexts, they question if there is something particular about neighbourhood as a category of socio-spatial configuration by looking at it through its absence in a certain socio-historical context. In doing so, they explore the potential the concept holds for understanding places like Eldorado Park.

Stories: neighbourhoods as imagined and narrated entities

The title of the Istanbul Biennial in 2017 was *A Good Neighbour.*[1] The curators argued that the Biennial 'will deal with multiple notions of home and neighbourhoods', exploring 'neighbourhood as a micro-universe exemplifying some of the challenges we face in terms of coexistence today'. In collaboration with the news website T24, the Biennial commissioned several writers from different disciplines to write short essays about a good neighbour. The Biennial also published a 'story book' titled *A Good Neighbour—Stories* bringing together personal stories and memories about homes, neighbours, and neighbourhoods. With incorporating these written outcomes into a visual arts exhibition, what the curators stressed was the narrative quality of neighbourhoods. While exploring different meanings and potentials of being neighbours, they overtly encouraged their audience to approach neighbourhoods as stories.

The second part of the book, bringing together neighbourhoods from Istanbul, Jerusalem, and Alexandria, also follows this conviction and focus on the presentation of neighbourhoods as *narrated* places—with imagined/denied pasts and futures, recounted/silenced memories and dreams, reproduced/ forgotten glories and squalor. The chapters in this part analyse how stories, collective memory(ies), and (un)official processes of naming, identifying, and idealising make a certain locality a neighbourhood.

There is, without doubt, serious contestation of the 'true story' and the legitimate story-teller. Claims are almost always met with counterclaims, which then lead to serious contestations of the place. The chapters in this part trace the plurality and alternating nature of these in their effort to represent a neighbourhood. In Annegret Roelcke's contribution, 'Two tales of a neighbourhood: Eyüp as a stage for the Ottoman conquest and Turkish War of Independence', for example, we see at least two competing narratives about the old Istanbul neighbourhood of Eyüp: one relying primarily on an imagined Islamic-Ottoman identity, and the other on a secular and ethnic Turkish nationalist identity of the place. Roelcke notes the 'palimpsest' nature of the place embodying traces of 'many different times and histories' (Huyssen 2003, 94).

Another significant case on the interwoven nature of contested material and narrative formations of neighbourhoods is provided by Johannes Becker, in his chapter entitled 'Past neighbourhoods: Palestinians and Jerusalem's "enlarged Jewish Quarter"'. With the backdrop of destruction of the neighbourhoods in the area of today's enlarged Jewish Quarter in Jerusalem and the eviction of most of the Palestinian residents, Becker brings to light the parallel processes of oblivion and disappearance. He emphasises that the 'life stories of neighbours' are intrinsically bound together with the 'life stories of neighbourhoods'. As much as the enlarged Jewish Quarter has lost its Palestinian residents and identity, it has also faded in the memories of its evicted and displaced former residents. And, even more strikingly, for the Palestinians who stayed, the quarter does not feel like a neighbourhood anymore and does not appear in the narratives of its inhabitants as such. With this insightful observation Becker suggests that neighbourhoods are lived and re-lived through the stories and memories of the people who remember and tell them.

Given the importance of immaterial processes (such as remembering, naming, narrating, longing, nostalgia, hoping, and dreaming) in the making and unmaking of neighbourhoods, the chapters also question the nostalgic and idyllic characterisation of neighbourhoods as consumable properties. Nostalgic claims about neighbourhoods come together with the trends of city branding, the demands of the tourism industry, as well as the market forces that turn cities into consumable properties and local relations into marketable experiences. Roelcke's chapter shows that differing claims about a place find expression in the interpretation of urban renewal projects and local heritage production in the neighbourhood. Social construction(s) and different constellations of Eyüp's imagined boundaries and the dominant stories of the neighbourhood determine 'heritage politics', municipal projects, local initiatives of musealisation, and tourism.

The story-telling dimension is quite pertinent in Samuli Schielke's 'Where is Alexandria? Myths of the city and the anti-city after cosmopolitanism', which focuses on the narratives of the city (and its neighbourhoods) created and reproduced by contemporary writers and literary circles in Alexandria. Schielke follows the traces of an ambiguous nostalgia for a bygone cosmopolitan/ colonial era, supposedly characterised by 'coexistence' and 'diversity', while the city is currently undergoing rapid transformation, demolition, and expansion. As the author notes, this narrative of nostalgia is a common feature of several cities in the Global South, given that many of them have been transformed into vast conglomerates within the past few decades, almost losing touch with their 'former selves'. Referring to a history of cosmopolitanism or a 'culture of coexistence', specific neighbourhoods are singled out in narratives as the real city, while others are excluded as the anti-city. The chapter also brings the city back into the discussion and reminds us that neighbourhoods are not closed entities. As Veena Das puts it, the differences among neighbourhoods are not unrelated to the way a particular neighbourhood is anchored to the city (Das 2017).

Yair Wallach's chapter in the part, 'Jerusalem's lost heart: the rise and fall of the late Ottoman city centre', approaches the cosmopolitan past and story-telling from the reverse side, that of denial, silencing, and oblivion. The chapter is about a forgotten neighbourhood, the town centre around Jaffa Gate, that emerged in the 1880s and embodied late-Ottoman notions of non-sectarian civic modernity, technological progress, and urban development. Wallach brings to light arbitrarily created administrative boundaries, spatial features/divisions, and planning/destruction of neighbourhoods, all of which have been major governmental preoccupations of colonial rulers and postcolonial nation-states alike. Some of these endeavours involved complete erasure of certain places, both from the physical reality of the city and from history books. Wallach delineates the physical destruction of the city centre by consecutive British and Israeli occupations, which also translated into its erasure from cultural memory. As this past 'neighbourhood' represented an 'undivided' Jerusalem that is not defined by segregation, it no longer served dominant nationalist narratives. In that sense, national, colonial, religious, and sectarian (hi)stories of nations and communities also define neighbourhoods.

Understanding the conflictual nature of the immaterial processes that make up neighbourhoods allows us to move from differed imaginations of neighbourhoods to the question of physical proximity, leading to either intimacy or conflict and violence between neighbours—themes that are explored in depth in the next part.

Intimacies: neighbourhoods as sources and objects of claim-making

In 2018 in Berlin, famous urban scholar AbdouMaliq Simone gave a keynote lecture in a panel entitled 'Right to the City'.[2] While speaking about how living arrangements and spatial geometries change for those who are involved in dissembling computers in the cities of South Asia, he refrained from using the terms 'neighbour' and 'neighbourhood', preferring 'residents' and 'residential areas' instead. When asked afterwards why he never used these terms, he talked about the superblocks, massive examples of the kind of so-called affordable, vertical housing that is becoming the norm in innumerable cities of the Global South, that seem to create an intensive homogenisation of everyday social life. These new residential complexes, he said, lacked and left no room for the intimacy that the concepts of 'neighbourhood' or 'neighbour' connote.

This part unpacks this connotation and the assumptions it implicates. Being neighbours is certainly a matter of proximity. Yet is it also and always a matter of intimacy? Earlier classics on urbanism explored this very same question at the level of the cities that were booming at the turn of the last century. Georg Simmel ([1905] 2012) turned his curious observation, that the metropolis had physically brought people almost within a touching distance of each other while creating insuperable social distances between them, into an insightful analysis and suggested that we city-dwellers develop a certain attitude to survive when surrounded by innumerable people whom we cannot possibly get to know. This attitude, blasé, immediately became one of the defining (and hugely famous) characteristics of urban environments.

While cities were identified with anonymity, apathy, and indifference between citizens, except a basic civility (*urbanitas* as Lewis Mumford (1937) calls it with reference to John Stow), neighbourhoods were often approached as havens of face-to-face encounters, intimate knowledge, and residence-based solidarity. Despite being urban formations and often seen as sub-categories to cities, they are also positioned in contrast. And not only on account of the social relations they lead to. Neighbourhoods are also positioned in contrast to cities in terms of their size and accessibility. The distances in a neighbourhood can be traversed on foot, while commuting in cities often requires more complex transportation arrangements, as the distances are too great to walk. With reference to these two major differences, the New Urbanism school in architecture goes so far as to suggest that 'a single neighbourhood standing free in the landscape is a village' (Duany and Plater-Zyberk 1994, 17). New Urbanists are not alone in seeing neighbourhoods as urban villages. Their vision of the neighbourhood with a 'human scale' (Duany and Plater-Zyberk 1994) material environment and intimate social networks is often shared by the nostalgic outcries for lost neighbourhoods.

The contributions in this book refer to this particular understanding of the neighbourhood as the locus of intimate relations, too. However, they do so only to unsettle it. They take 'both the constraints and the companionship of spatial solidarity' (Herzfeld 1991, 88) seriously and place them under analytical scrutiny to develop a spatialised understanding of the formation of networks, the creation of discourses, and circulation of affects. Through this focus, they resist conflating the physical and the human aspects of an urban space. In another words, while recognising the significance of proximity, they do not use it as a synonym for positively valued intimacy. Proximity in the urban space appears as prone to create conflicts, animosities, and cleavages as it does to produce solidarities and genial sociabilities.

In this part, the chapters focus on both affective ties, neighbourliness, and personal relations and the conflictual moments of claims-making that put neighbourhood identity to the test. Aatina Nasir Malik's chapter, 'Violence, temporality, and sociality: the case of a Kashmiri neighbourhood', based on ethnographic research in Nowhatta, a downtown neighbourhood of Srinagar, goes beyond the narrative that presents Nowhatta as the cradle of Kashmiri insurgency. Despite the stereotypical descriptions of the neighbourhood in mainstream media as either a hub of violent youngsters or a victim of state violence, Malik offers an exploration of how violence and everyday life interact, and even blend into each other. Focusing on the imposition of curfews as the main form of state violence, the chapter argues that not everyone in the neighbourhood is affected or involved in the same way. Nowhatta, in the end, is a microcosm where the border conflict between occupying India and interventionist Pakistan is lived and counterposed with the demands of insurgents seeking independence and the people whose main concern is to have a normal life.

Malik's chapter is not the only one in this part showing how residents of a certain locality, despite the close proximity with which they live to one another, do not always become neighbours. What initiates neighbourly relations are encounters of various kinds and relations that evolve from these encounters. Neighbourhoods are enacted compositions of social relations in a particular locus. They are, as Doreen Massey argues for places, indeed constellations of social relations (1994). However, neighbourhood relations are not always amiable, as implied by the term 'neighbourly'; rather, they often involve contestation, conflict, and violence. An increasing number of works on conviviality and neighbourhood violence suggest that heterogeneity, conflict, and dividedness are quite frequent despite the idyllic nature of the term 'neighbour'. The everyday violence that takes place in the intimate space of neighbourhoods, directed at those who are personally known to the perpetrators, is of common occurrence all around the globe. While age-old neighbours may turn into eternal enemies through mass violence, new migrants may also face similarly unwelcoming attitudes, alongside hospitality and solidarity. Xenophobia and hostility can be defining marks of a neighbourhood as much as welcoming and openness, especially in our contemporary age of migration and movement.

Hilal Alkan's chapter, 'Syrian migration and logics of alterity in an Istanbul neighbourhood', and Derya Özkaya's chapter, 'Negotiating solidarity and conflict in and beyond the neighbourhood: the case of Gülsuyu-Gülensu, Istanbul', both explore the processes that turn residents of a locality into neighbours and neighbours into enemies. Hilal Alkan considers how a logic of alterity is played out in the neighbourhood of Kazım Karabekir, which is now home to an increasing number of Syrian migrants. Residents of the neighbourhood respond to the challenge of encountering and living with 'strangers' within a spectrum that ranges from forming solidaristic support networks to hostile harassment of their new Syrian neighbours. The daily mechanisms of these strategies lead us to question the conditions and possibilities of neighbourliness, particularly because such conditions are very much intertwined with national and international politics.

As the leading regional socio-political issue, the Syrian war finds a place, also, in the lives of Gülsuyu-Gülensu residents, as illustrated in Özkaya's chapter. But the challenge they are facing is not about how to respond to their new neighbours, but the active involvement of the inhabitants of the neighbourhood with the war. The war is literally being re-placed and restaged in the neighbourhood at the moment, with many youths actively taking up arms in Syria and then returning to Gülsuyu-Gülensu. Hence, it is a question of how to remain neighbours with the person one fights against in a different context. This question has deepened the existing violent political rift between the two sides of the neighbourhood and made the formerly strong solidarity against urban transformation almost obsolete. The Gülsuyu-Gülensu case shows us how fragile neighbourhood relations can be in the face of competing political orientations at the intersection of national and international developments, and how violence and the threat of violence may unmake a neighbourhood formerly known for its close-knit relations.

Certainly, conflicts in the neighbourhoods are not simply local reflections of larger incidents. All of the chapters in this part show, through rich ethnographic accounts, how these different scales entangle and each neighbourhood's own power configurations tap into these entanglements. At the neighbourhood level, such conflicts often find expression as claim-making, and a major way to claim a place is to use it, to perform daily actions, and to physically occupy the space. Such claims-making involves contestations, not only based on ideology, but also (and very much) on gender, ethnicity, class, and morality. A prevalent example of this can be found in Urszula Ewa Woźniak's chapter, 'Urban tectonics and lifestyles in motion: affective and spatial negotiations of belonging in Tophane, Istanbul'. Woźniak illustrates the territorialised conflicts over lifestyles in a central Istanbul neighbourhood, where residents are particularly afraid of losing their neighbourhood to pacing gentrification. Their anxieties are expressed in a hostile attitude towards the neighbourhoods that surround Tophane and towards the newcomers who settle or establish businesses in the neighbourhood. As much as these mundane conflicts are linked to macro-level political and socio-economic tensions, they create powerful and sometimes violent clashes. Yet, they are often framed as clashes over the prevailing moral order of the neighbourhood and the social identity established around this particular mode of morality.

In the last chapter of this part, 'The Basij of neighbourhood: techniques of government and local sociality in Bandar Abbas', Ahmad Moradi examines a neighbourhood in Bandar Abbas to decipher the complexities involved in establishing the particular moral order sponsored by the Iranian state via the local Basij bases. Moradi shows that these bases are tangible centres for the exertion of governmental power over neighbourhoods, yet he also highlights the ongoing compromises that the Basijis make as they acknowledge neighbourhood relations. By examining lines of disputes and conflicts embedded in everyday life, as well as the intimate relations of neighbours, Moradi emphasises the need to examine on-the-ground understandings of 'techniques of government' and 'social relations'—and of the flexibility of the boundaries between them. As he aptly illustrates, neighbourhood provides the ideal scale to achieve this.

Notes

1 15th Istanbul Biennial, *A Good Neighbour*, 16 Sept.—12 Nov. 2017. Curated by Michael Elmgreen and Ingar Dragset. http://15b.iksv.org/agoodneighbour. Accessed 22 September 2019.
2 HAU Hebbel am Ufer, 23 June 2018. https://produktionshaeuser.de/wp-content/uploads/2018/06/BIP_Programmheft_Web.pdf?x68854. Accessed 22 September 2019.

References

Alimia, Sanaa, Andre Chappatte, Ulrike Freitag, and Nora Lafi. 2018. *In Search of Urbanity.* ZMO Programmatic Texts, 12. Berlin: Leibniz-Zentrum Moderner Orient.

Chaskin, Robert. J. 1997. 'Perspectives on Neighborhood and Community: A Review of the Literature.' *Social Service Review*, 71(4): 521–547.

Das, Veena. 2017. 'Companionable Thinking.' *Medicine Anthropology Theory*, 4(3): 191–203.

De Vito, Christian G., and Anne Gerritsen. (eds.). 2017. *Micro-Spatial Histories of Global Labour.* Cham: Palgrave Macmillan.

Döring, Jörg, and Tristan Thielmann. (eds.). 2015. *Spatial Turn: Das Raumparadigma in den Kultur-und Sozialwissenschaften.* Bielefeld: Transcript Verlag.

Duany, Andres, and Elizabeth Plater-Zyberk. 1994. 'The Neighbourhood, the District and the Corridor.' *The New Urbanism: Toward an Architecture of Community*, P. Katz, V. J. Scully and T. W. Bressi (eds.). New York: McGraw-Hill: 17–20.

Epple, A. 2012. *The Global, the Transnational and the Subaltern. Beyond Methodological Nationalism: Research Methodologies for Cross-Border Studies*, A. Amelina et al. (eds.). New York: Routledge.

Friedmann, John. 2010. 'Place and Place-Making in Cities: A Global Perspective.' *Planning Theory & Practice*, 11(2): 149–165.

Herzfeld, Michael. 1991. *A Place in History: Social and Monumental Time in a Cretan Town.* Princeton: Princeton University Press.

Huyssen, Andreas. 2003. *Present Pasts: Urban Palimpsests and the Politics of Memory.* Palo Alto: Stanford University Press.

Jacobs, Jane M. 2012. 'Commentary—Comparing Comparative Urbanisms.' *Urban Geography*, 33(6): 904–914.

Massey, Doreen. 1994. *Space, Place and Gender.* Cambridge: Polity Press.

Mumford, Lewis. 1937. 'What Is a City.' *Architectural Record*, 82(5): 59–62.

Rechtman, Richard. 2017. 'From an Ethnography of the Everyday to Writing Echoes of Suffering.' *Medicine Anthropology Theory*, 4(3), Special Section: On Affliction: 130–142.

Robinson, Jennifer. 2006. *Ordinary Cities: Between Modernity and Development.* New York: Routledge.

Robinson, Jennifer. 2016. 'Comparative Urbanism: New Geographies and Cultures of Theorizing the Urban.' *International Journal of Urban and Regional Research*, 40(1): 187–199.

Roy, Ananya. 2011. 'Slumdog Cities: Rethinking Subaltern Urbanism.' *International Journal of Urban and Regional Research*, 35(2): 223–238.

Simmel, Georg. [1905] 2012. 'The Metropolis and Mental Life.' *The Urban Sociology Reader*, J. Lin and C. Mele (eds.). New York: Routledge: 37–45.

Warf, Barney, and Santa Arias. (eds.). 2008. *The Spatial Turn: Interdisciplinary Perspectives.* New York: Routledge.

Wirth, Louis. 1938. 'Urbanism as a Way of Life.' *American Journal of Sociology*, 44 (1): 1–24.

Part I

Borders

Material, temporal and conceptual
boundaries of neighbourhoods

1 What makes a township a neighbourhood?

The case of Eldorado Park, Johannesburg

Alex Wafer and Kholofelo Rameetse

Introduction

In this chapter we consider Eldorado Park, a primarily residential and commuter area roughly 20 km to the south west of Johannesburg, and surrounded by suburban commuter sprawl, from the perspective of *neighbourhood*. Unlike some of the other contributions in this book, which proceed from the relative self-evidence of particular sub-metropolitan urban areas as neighbourhoods, in this chapter we use the case of Eldorado Park to ask the question: what actually constitutes a neighbourhood? This may seem at first glance a rather pedantic semantic exercise: with a population of just over 60,000 people (StatsSA 2011), with local-scale commercial and public amenities (Eldorado Park has several schools, a clinic, and a public library), relatively clearly defined geographic boundaries, and a strong sense of locational and community identity among many of its residents, Eldorado Park would seem to constitute a scale and context that encompasses the everyday life-worlds of its residents. And yet the term 'neighbourhood' seems somewhat out of place in reference to Eldorado Park.

In part, this is simply a matter of semantics. The awkwardness of the term in reference to Eldorado Park reflects the peculiarities of a South African vernacular (and urban policy language) that continues to refer to (and therefore imagine, we would argue) urban space in terms of older apartheid-era categories: i.e. suburbs (primarily former white group areas), townships (former black and 'coloured' group areas) and more recently 'communities' (a non-specific term that recalls an anti-apartheid politics of 'community-based protest', but which now variously encompasses sub-township or street-level geographies, as well as informal settlement areas, post-apartheid public housing projects, and the more recent trend towards gated private developments). This is not to say that a person using the term 'neighbourhood' in reference to Eldorado Park would be misunderstood. In so far as they were referring to the area as one among many sub-metropolitan scale residential areas with minimally definable boundaries and internal self-identification, the term would be appropriate. But at a more substantial level, we wonder here whether neighbourhood could be simply another moniker for a township or suburb, for example, or whether these objects of analysis are in fact distinct? And if they are distinct, do they produce different imaginaries of being and belonging?

The contention then becomes that the out-of-placeness of the term 'neighbourhood' in reference to Eldorado Park reflects more than just semantic preference, and that the socio-spatial categories that do pertain (i.e. township, community, etc.) reflect and reproduce the ways in which South African cities continue to be inhabited, governed, and imagined. That is to say, perhaps we seldom use the term 'neighbourhood' because neighbourhoods don't really exist in South African cities. If this is indeed the case, then the question that emerges is what is lost or gained by this fact? Is there something about neighbourhood as a socio-spatial category that South African cities lack? Alternatively, we might also conclude that the categories that do frame our urban imaginaries blind us to the ways in which ordinary people do in fact organise their everyday life-worlds in the urban spaces and at the local scales that they inhabit. If 'neighbourhood' is understood as more than simply one among many interchangeable terms that describe sub-municipal demarcations – in other words, if there is something particular about neighbourhood as a category of socio-spatial configuration – then could South African urban studies gain something by thinking through this geography? And conversely, could a literature that takes neighbourhoods to be largely self-evident benefit from exploring precisely such a sub-municipal configuration not conventionally thought about in these terms?

In this chapter we explore this proposition. As such, we request that the reader approach the chapter in the spirit of intellectual speculation with which it is offered. We did not begin our research in Eldorado Park with the intention of thinking of this place – simultaneously banal and crushing of the human spirit, as well as tragicomic and life-affirming – through the lens of neighbourhood. Perhaps to some degree our research aim was itself inflected by the very conceptual bias that we identify above: seeing the South African city laid out before us in a recognisable form from the perspective of a researcher. Much urban research begins by identifying an empirical problem: in our case, tensions emerging around social housing between residents of Eldorado Park, the adjacent former black township of Kliptown, and the more recent informal settlement that has grown between them. Of ongoing significance is the legacy of apartheid-era racial identification that was written into the spatial organisation of South African cities, and that is often subtly (and sometimes explicitly) invoked in attempts by residents to access the post-apartheid state. Our primary interest has been in the ways that the everyday materiality of urban life in South African cities intersects with aspirations of political subjectivity: how people use apartheid-era infrastructure to make claims of the post-apartheid state, for example; or how communal taps in informal settlements become the basis for articulations of more substantive citizenship. However, as we have come to know Eldorado Park as exceeding the archetypes of apartheid urban planning or of post-apartheid urban abandonment, we have also become more attuned to the ways in which its residents build durable and meaningful lives in its inadequate but nevertheless given material presence. Whether this counts as neighbourhood, or the extent to which the concept of neighbourhood is a useful analytical tool, will be explored in the sections that follow.

We approach the concept of neighbourhood through three key propositions: (1) as a socio-spatial configuration within which people's life-worlds are located, and sometimes corresponding to a specific place or 'heart' that lends broader meaning to the whole (Meegan and Mitchell 2001); (2) as a socio-spatial scale at which local residents feel a shared experience (Forrest and Kearns 2001); and (3) as a socio-spatial configuration that reinforces a place-based (and sometimes exclusionary) identity (MacAllister et al. 2001). Whereas the 'heart' of a neighbourhood is often thought of in terms of a marketplace, public square, or even a main street, we ask whether a police station might be considered a meaningful place through which life-worlds are produced. This is not simply because the police station lies at the geographic centre of Eldorado Park, but because it is for many residents the public space around which the most intimate and personal elements of their everyday lives are organised. Designed originally as a space of surveillance, in the post-apartheid context it has been remade as a space through which the absent state is imagined in both its disciplinary and pastoral form. We subsequently consider Eldorado Park as a socio-spatial scale of shared experience through the concept of compression, through which the over-saturation of public resources creates an experience of manufactured scarcity. Finally, we consider how often-exclusionary identity is reproduced by the material context of Eldorado Park, which we consider using the concept of abandonment. Here we reference Povinelli's (2011) notion of abandonment to show how apartheid racial categories are invoked in the post-apartheid context as the basis for claims of (unrealised) citizenship. But before considering these three propositions, in the following section we re-pose the question: why think of Eldorado Park through the lens of neighbourhood at all?

Neighbours without neighbourhoods

Eldorado Park was established as a racially designated 'coloured' township by the apartheid government in the 1950s (Brown 2013). Situated around 20 km south-west of the city, it was part of the larger South Western Townships (popularly called Soweto). The racial classification of 'coloured' has a complex and difficult history in the South African context: under the apartheid racial taxonomies upon which the socio-spatial order of apartheid was organised, 'coloured' referred to people of mixed-race heritage, primarily the illegitimate progeny of European settler-colonists and slaves or other indentured communities. For example, in much of what are now the provinces of the Western Cape and the Northern Cape (formerly the Cape Colony, and then the Cape Province under apartheid), between 40 and 50 per cent of the population self-identify as 'coloured' (StatSA 2011). Note that these racial categories are still officially recorded in South Africa, twenty-five years after apartheid, although a person is not obliged to declare a racial identity. Notwithstanding the attempts at undoing the racialised society, the geography of race and racial separation persists. In Johannesburg, 'coloured' is a minority racial category, and Eldorado Park remains the area with the highest concentration of people who identify as 'coloured': in the 2011 census over 80 per cent of residents of Eldorado Park identified as 'coloured'.

Under apartheid, 'coloured' racial classification meant exclusion from meaningful political and economic participation, although in the peculiar racial matrix, 'coloured' (as well as Indian, in reference to those people who had arrived under colonial rule from the Indian sub-continent as indentured labourers) was nominally slightly more privileged than the racial classification of African. In practice, 'coloured' communities have suffered social, political, and economic exclusion comparable to other non-white groups, and the legacies of 'coloured' exceptionalism have meant that in the post-apartheid context, racially identifying 'coloured' people internalise a strong sense of minority persecution (Wicomb 1998). It is not the purpose or the place of this chapter to interrogate what remains a highly complex question of racial identity and inequality in South Africa. It is simply to give the reader some very general context to help them understand the specificity of Eldorado Park. As for Eldorado Park itself, as a socio-spatial and material configuration, the township today exhibits many of the characteristics that defined its establishment. It remains largely racially homogeneous, economically marginalised, and deeply implicated in the criminal-penal complex.

As intimated in the Introduction, there are primarily two reasons why one might hesitate to refer to Eldorado Park as a neighbourhood. The first has to do with the neighbourhood as an object of social research, specifically in the South African urban studies tradition. This is not to say that neighbourhoods do not exist in South Africa, but rather, it is recognition of the conceptual register (and therefore the plane of imagination) through which South African cities have historically been researched and theorised. Unsurprisingly, urban studies in South Africa has been preoccupied with the effects and legacies of racial segregation. This was initially theorised through materialist attempts at explaining an apartheid political-economy based on cheap migrant labour (see e.g. Wolpe 1972). The South African city was largely dismissed as an extension of the colonial city: i.e. a space of racialised exclusion and administrative and economic power. It was the rural areas, and their relationship to the mining economy, that were of most interest. To the extent that South African cities emerged as objects of academic interest in their own right, in this tradition they were seen largely as the spatial logic of a particular configuration of capital. Gelb's (1991) definition of 'racial Fordism' remains a powerful theory for not only the macro-structure of the apartheid political economy, but also the development of the archetypal apartheid city: i.e. a model of urban development that closely resembles twentieth-century North American cities, albeit with an explicitly racial-segregationist logic superimposed.

Against the dominance of this materialist tradition in South African social sciences in the 1970s and 1980s, what might be called an 'urban' turn emerged in the 1980s and 1990s. One key intervention was a volume published in 1992 titled *The Apartheid City and Beyond* (Smith reprinted 2003). This was the sequel to a volume published a decade earlier, although whereas both volumes were primarily about urban life under apartheid, this second volume was explicitly concerned with the city as both site and object of political contestation. Again, this is unsurprising. By the 1980s the apartheid system was in crisis largely because of changes in the urban areas: declining real wages, inadequate basic services and a growing housing crisis in designated black urban areas, and an emerging youth political movement that rejected the acquiescence of their parents together precipitated a decade of urban-based violence that cost thousands of lives and defeated the apartheid regime. In this context, it was the township that emerged as the object of academic fascination (Robinson 1996).

While there is arguably no intrinsic geographic distinction between township and neighbourhood – both refer to a sub-municipal residential area – the governmental (and subversive) logics of such socio-spatial configurations were of primary interest. Townships were of political and academic interest because of their relationship to the larger urban system of apartheid, rather than as particular urban spaces with unique and meaningful characteristics. In fact, townships were critiqued precisely because of their planned lack of uniqueness: for the drudgery and monotony of their material form, which seemed designed to deliberately exclude the possibility of meaningful everyday life beyond labour reproduction. As such, the scalar imaginaries of urban politics in South Africa have always been the very local (the community or the street committee), on the one hand, and the metropolitan scale on the other. A prominent slogan of the civics (the community-based organisations that contested urban apartheid in the 1980s) was 'one city, one tax base' – a reference to the metropolitan scale rather than the sub-metropolitan scale as political objective (Wafer 2012).

More specifically, the neighbourhood is not a traditional spatial scale of analysis in much of South African urban studies. It is not to say that we do not have neighbourhoods in South Africa, but there is nevertheless a sense in which the conceptual language dictates the possibilities of imagination. More conventionally, South African urban scholars have studied the apartheid city as a spatially and functionally segregated urban system. In this matrix, the units of analysis are not individual neighbourhoods, but rather the '(white) suburbs' as a generalised term to refer to those areas of historical and geographical privilege (Benit-Gbaffou et al. 2012; Dirsuweit and Wafer 2005), with the township as a referent to those racially designated areas of exclusion and segregation. Despite the ending of apartheid, and more significantly, the radical spatial realignment of the urban economy in the past three decades, these generalised terms continue to have valance in South African urban scholarship and policy making. The Breaking New Ground policy that defines the South African housing programme – one of the key vehicles through which poor people encounter the state – mentions 'the community' or 'communities' on 37 occasions throughout the 40-page document. It mentions townships on six occasions, mostly in reference to historical public housing arrangements. There are two references to neighbourhoods (in the same paragraph) in regard to historically middle-class white group areas (Government of South Africa 2004). Given that the emphasis of the programme is on developing sustainable settlements (as opposed to the earlier post-apartheid housing programme, which was driven by numbers, see e.g. Ballard and Rubin 2017), the lack of reference to the neighbourhood is remarkable.

It would not be unfair to say that this lack of attention to other forms of socio-spatial organisation in South African urban scholarship has been somewhat of an oversight, in which the nuances of the local and the particular have been overlooked in place of the perceived need to generalise about the South African urban condition.

The second reason why one might hesitate to refer to Eldorado Park as a neighbourhood is more subjective. As an exemplar of the apartheid designation of township, the area lacks any of the material and affective qualities that might be regarded as neighbourhood (Swilling and Annecke 2006). In this regard, the apartheid urban project was so successful that the possibility of neighbourhoods – which have distinct local identities and which contain the life-worlds of residents (see e.g. Meegan and Mitchell 2001) – has been thoroughly erased from the urban landscape. In its place are the traumatised remnants of a racialised socio-spatial order: i.e. the displaced and disfigured members of 'the community'. Community is a non-specific indicator of social and political agency which mostly implies a kind of proto-subject of post-apartheid transcendence (Chipkin 2007). 'Communities' appear to exist everywhere and nowhere in the contemporary South African city.

Obviously the distinction is subtle, since communities are also defined by their connection to particular places. But in the context of South African urban dynamics, place here is less about the specifics of geography in any durable meaningful sense and more about the generics of racial classification (and the geography they imply). The implication, in the post-apartheid context, is that communities rather than neighbourhoods are the subjects of state intervention. In every ministry, at national and provincial levels of government, it is 'the community' that is invoked as the ur-form of the political subject – even though the 'citizen' is the subject enshrined in the post-apartheid constitution. The effect is that government interventions are frequently generic and homogenous, directed less towards enriching the possibilities for the individual-as-citizen to realise her rights through the city, and more towards realising collective demands articulated in terms of quotas. It is no surprise that much post-apartheid social housing, for example, is frequently criticised for being no better than apartheid township housing (Haferburg 2013; Lemanski 2009).

Though not articulated in the same language, Partha Chatterjee's (2011) distinction between civic and political society might be applicable here. If civil society implies particular expectations of the urban form, including but also exceeding basic service delivery (i.e. civic spaces, libraries, museums, public transport, etc.), then the standard response to community protests for state services is not about embedding a set of urban infrastructures around which a sense of place and belonging can emerge, but about providing the most minimal levels of access to the greatest number of beneficiaries. For example, public housing is frequently built on cheap greenfield sites on the urban peripheries and residents of informal settlements are often relocated to these sites, rather than more innovative attempts at allowing existing proto-neighbourhoods to flourish (Charlton 2013). This is a very powerful socio-spatial imaginary – of community as a group of people with common cause rather than the neighbourhood as a material and affective geography – and has, we argue, remained dominant in the South African urban imaginary.

It also means, not unexpectedly, that urban dynamics are frequently rendered conflictual and exclusionary. Whereas a neighbourhood might contain a range of resources that a resident can draw upon, communities are frequently invoked as ways of capturing networks of patronage. That is, communities represent voting blocs that can be mobilised (as opposed to neighbourhoods which may have political tendencies one way or the other, but are first and foremost households of citizens at least in principle with pluralistic political identities). The cycles of xenophobic violence that flare up in the poor estuarial spaces of South Africa's large cities every few years are suggestive of the very populist politics that a discourse of community can support (Misago 2019). Similarly, although at the other end of the socio-economic spectrum, the discourse of community has emboldened residents in areas of extreme privilege to effectively exclude themselves from the political dynamics of the city, in the form of gated communities (Dirsuweit and Wafer 2005). This is not confined to the South African context. Aptekar (2017) has shown how the strong sense of neighbourhood in places like Astoria, in New York, forged over multiple generations and in spite of cycles of changing demographics (e.g. from a predominantly Greek to a predominantly Turkish to a predominantly East European neighbourhood) can be undermined by the immigration of 'gentrifiers' whose routines and digital-cosmopolitan aspirations effectively separate them from the neighbourhood as a site and object of affective implication, even as they reside within it geographically.

So why might we want to look at Eldorado Park through the lens of neighbourhood? One reason is to think through the alternative socio-spatial scales at which people organise their everyday lives, and which may not be captured in the more dominant socio-spatial categories of South African urban studies. The neighbourhood presents a scale and an empirical object that has not been analysed much in the South African context. Certainly, thinking about Eldorado Park through its affective and everyday meanings, through the relations between people and places at the local scale, rather than primarily through its generic characteristics vis-à-vis the urban system offers something potentially new to South African urban scholarship. This would hardly be the first study to do just that in areas that are conventionally considered townships, although to our knowledge there are no studies that have focused on Eldorado Park. Indeed, most published studies on Eldorado Park concern the public health, public violence, and anti-social behaviour that characterise the popular imaginary of the township. On the other hand, the reader might find it interesting to take a seemingly self-evident concept such as neighbourhood, with all its attendant significations, and challenge its usefulness as an explanatory term. If nothing else, we hope that this chapter serves this function. But finally, we are compelled to consider Eldorado Park through the lens of neighbourhood as an intervention rather than simply an exercise in academic contemplation. In the South African context, the term 'neighbourhood' largely connotes loss or lack of urban life that was (in some mythical past) something that was at the scale of individuals and their everyday social relationships. Whether these are tragic and violent, or comforting and hopeful, they are nevertheless rich and meaningful. A concept such as neighbourhood is of course not required to capture the meaningful life-worlds of ordinary people: perhaps all such terms are insufficient. But to the extent that we might imagine more hopeful urban futures in South Africa, thinking of the familiar through a different lens is an important scholarly and ethical gesture. The remainder of the chapter is directed to this task.

Surveillance

Eldorado Park is the archetype of apartheid urban planning. In plain view, Eldorado Park illustrates the intersection of a panoptic logic with the influence of a settler-colonial obsession with the suburban garden city (Findley and Ogbu 2011). The township is perfectly cut off from the rest of the city. To the south, Eldorado Park is bordered by the N12 highway, to the east, by the R553 'Golden Highway', which separates it from neighbouring Devland Township (now mostly referred to as Freedom Park). A large tract of empty land on either side of the R553 was the original buffer zone between Devland and Eldorado Park, and this has subsequently been developed by private housing developers. This has been the cause of recent conflict between residents of Eldorado Park and Freedom Park, something that we will discuss below. To the north, Eldorado Park is bordered by a vast uninhabited terrain that includes the Goudkoppies municipal landfill and then the municipal water treatment plant. A small patch to the north-west was carved out for the erstwhile Soweto Country Club, a golf course established in 1974 for black South Africans, now largely abandoned. To the west, Eldorado Park is separated from Kliptown by the Klipspruit wetland, much of which has been informally settled in the post-apartheid era.

The township itself consists of tiny free-standing bungalows, each in the centre of a small dusty yard: a mean and pale version of the suburban imaginary. The streets are organised as a series of meandering crescents, all ultimately looping back onto Main Road: like ribs along a central spine that runs along an east–west axis. Despite its centrality as an organising principle in the township layout, Main Road is not the central promenade along which economic activity and social life happens. Some large- and medium-scale retail exists along this route, accessible via car, but for the most part Main Road is an inhospitable non-place (Augé 2008): a four-lane dual carriageway with wide shoulders and constant through traffic, while most of the everyday life of the area turns its back to it. Main Road was never designed as a place of public street life. Its primary purpose is vehicular access into and out of the township: it remains the only access route. During apartheid, this served the purpose of enclosure and surveillance: it was an easy access and patrol route for armoured police vehicles. And while the warren of residential streets that branch off from Main Road may have provided cover for protestors fleeing police violence, especially in the 1980s, they would never escape the labyrinthine logic of the grand strategic plan: the township was designed to keep people locked inside. This logic pervades even today, both physically and psychologically, as Eldorado Park is widely regarded as something of a ghetto cut off from the rest of the city. The sporadic community protests that characterise Eldorado Park's public image usually begin with residents closing down Main Road; a political gesture that serves to shut the township down while also ironically speaking to its pre-existing isolation.

The Eldorado Park Police Station is situated at the midway point along Main Road, at the corner of Buckingham Avenue. In the original design logic, the police station was intended as the panoptic epicentre of what has sometimes been referred to as the geography of exception (Cooper-Knock 2017). It stands sentry in the midst of an architecture of dehumanising mundanity, buffered from the surrounding community by the grounds of the local secondary school to its south and the primary school behind it to the north. On its eastern flank is a vast vacant strip of no man's land, now partially occupied by a cheap supermarket. It is a fortified compound of bland, low-level, brick-faced buildings originally built in the mid 1960s and given a half-hearted upgrade in the late 1990s to make them slightly more hospitable. In the post-apartheid context, the South African Police Service (SAPS) has attempted to portray itself as community responsive and supportive, a changed institution from the instrument of state-sponsored violence and oppression during apartheid (Rauch 2000). In places like Eldorado Park, where rates of violent crime and community disaffection about service delivery are high (Mabasa 2012), residents remain deeply suspicious of the police. In many ways, the original function of the police station as the panoptic epicentre of the township remains intact. But as we will show, its role has also shifted to one of (albeit insufficient) pastoral work.

Our first introduction to Eldorado Park was via the police station when we accompanied a former Eldorado Park resident, with whom we had been working separately, to get information on the case of his daughter's unsolved murder. During several fruitless visits over the next several weeks, we learned very little about our colleague's case but a huge amount about everyday life in Eldorado Park. As with many over-subscribed public institutions in poor neighbourhoods in cities across the world, the police station waiting room was a fetid mixture of tense expectation and endless resignation. The stale air was thick with the odour of waiting bodies. Women and men of all ages, bent over with the exhaustion of poverty and despair, clutching old and worn documents, sat on battered institutional benches. Austere neon tubes buzzed and flickered in a slow but maddening rhythm, casting a ghastly pall over the whole ensemble. Despite the endless waiting across the counter, there always seemed little urgency in the back offices, as bored-looking police officers shuffled languidly through worn papers and documents. Such scenes undoubtedly occur in thousands of police stations across the world every day. But in Eldorado Park, we would learn that a far deeper co-dependence exists between the residents and the township–police–prison nexus – inherent, we argue, in the spatial organisation of the township.

Few families in Eldorado Park do not have some ongoing relationship with the police and prison system. It is a less than 15-minute drive (around 8 kilometres) from the Eldorado Park Police Station to Johannesburg Prison, popularly called Sun City. Many of the male members of families living in Eldorado Park are currently or have previously been incarcerated there. In our on-and-off research in Eldorado Park over the past several years, we have scarcely met anyone who has not been involved in this carcereal geography. Young people, especially young men, refer to Eldorado Park as a waiting room for the prison. At the epicentre of this geography is the police station itself. Residents remain incredibly mistrusting of the police, especially young people. It is via the police station that young men are processed upon arrest for any number of gang-related and anti-social behaviours; and it is here that mothers, grandmothers, and wives must come to bail their sons and husbands out – sometimes after having reported these same relatives for domestic violence. It is also here that residents must come to access more prosaic forms of social justice: signed affidavits that might stand up in court cases to access child support, or reporting faulty infrastructure, for example. The police officers know first-hand multiple generations of local families as well as their individual tragedies and everyday lives: whether they are at school, or employed, or involved in illegal activity or community activism. Police officers speak of the kinds of psychosocial care that their job demands: something most of them feel ill-equipped to adequately offer. In the absence of any other more tangible manifestations of the state, the police station serves this pastoral function.

Everyday life in Eldorado Park is constituted far more richly than by visits to the police station, of course: we do not have space here to explore the significance of churches and mosques, of which there are several dozen in Eldorado Park and which play an important role in the life of many residents. Schools also play an important role in community politics. But whereas many of these institutions constitute smaller communities of socio-spatial retreat, we would argue that the object of the police station is most intricately implicated in the everyday life-worlds of residents. Corbridge et al. (2005) are among a number of scholars to demonstrate the ways in which ordinary people encounter the state imaginary not through the formal rights and institutions of democracy, but more often through the personalised (and often patrimonial) relationships with local state functionaries. Read alongside the work of scholars such as Anna Secor (2007) who has highlighted the affective qualities of state institutions – the co-presence of the sacred and the profane in the imaginary of the state – it is possible to see in the police station something of the hope and despair that permeates everyday life in Eldorado Park.

Compression

In the previous section we discussed Eldorado Park in terms of what we called surveillance, meant here not so much as a constant vigilant watchfulness than as the implication of everyday life into police/prison/township geography. In other words, even the most intimate details of people's everyday lives are organised through the unintended logic of the police station in the absence of any other institutions of the state. We suggested that the police station represents in this context the less-than-ideal heart of the township – i.e. it is the organising principle that structures social life. If surveillance understood in this sense is the organising logic, then the edges or limits of Eldorado Park might be experienced through the concept of what we call compression.

In most South African cities the local scale of urban politics is preoccupied with the traditional networked infrastructures of the modernist city (Graham and Marvin 2002). This has everything to do with the particular history of unequal access to such services; one of the ways in which racialised segregation was inscribed into the urban form was through the unequal provision of what are called basic services in South Africa. Service delivery protests – i.e. protests about the slow pace of delivery of water and electricity connections to poor households – are a common feature of South African urban politics (Ballard et al. 2005). Where these traditional networked infrastructures do exist (in wealthy suburbs, but also in older parts of the city including township areas that are relatively well serviced compared to informal settlements), what we see is not so much a splintering of these urban infrastructures (as Graham and Marvin argue), but rather forces of compression.

Leslie Bank (2011) refers to these compressive forces as fracturing urbanism: i.e. the cracking and breaking of the city under compression. In other words, rather than (or at least in addition to) new technologies and economies driving a dissolution of the traditional compact urban form, and the consequent disarticulation of traditional infrastructural modes, many of the urban poor compress into those pockets of the city that provide access to albeit over-saturated networked infrastructures. In some parts of Johannesburg this compression has literally caused the collapse of the material form of the city (see e.g. Wilhelm-Solomon 2017).

These compressive forces are prevalent in many parts of the city from which capital and urban management have fled, such as parts of the inner-city and older industrial areas. But it is former township areas that have arguably undergone the greatest compression. Population densities in many former townships are some of the highest in the city, despite their low-scale, low-density design. The form through which this takes place is the phenomenon of the back-yard room (Lemanski 2009). Back-yard rooms are rooms (sometimes as rudimentary as sheds) built in the back yards of township houses (they are also increasingly common in the back yards of newer subsidy houses). Most back-yard rooms are built without any planning permission and in technical violation of multiple municipal by-laws.

Nevertheless, they are an urban typology that has emerged in the post-apartheid city as a bottom-up response to a housing crisis. In fact, back-yarding is an urban practice that perfectly complements the township form: the low-scale bungalow in the small dusty yard is a fusion of ideological histories and political-economic possibilities that has produced a post-apartheid urban landscape of unequal sub-urban sprawl. In such a context, township residents segregated under apartheid and marginalised in a post-apartheid spatial economy have been able to reclaim some degree of economic autonomy by renting out back-yard rooms. An aerial photograph of Eldorado Park reveals hundreds of these back-yard rooms (see Figure 1.1). The original township house is clearly identifiable: a repeated pattern formed by the beige oblong roofs with what appear to be chimneys (in fact, these are recently retro-fitted solar geysers) surrounded by an array of shapes and colours representing back-yard structures. Most appear small and rudimentary; while some are fairly substantial constructions. We documented as many as fifteen rooms built haphazardly and cheaply in a single tiny back yard – although most back yards have only two or three.

Figure 1.1 Aerial view of part of Eldorado Park (police station in upper right quadrant).
Source: Google Earth.

In a context where unemployment is higher than the urban average (Matjomane et al. 2018), backyarding provides a vital source of income for most homeowners in Eldorado Park, but it also contributes to urban compression. Sometimes this compression produces actual material fractures: infrastructures break, sewage pipes rupture, roads deteriorate. Sometimes the fractures emerge in the form of social conflict. Back-yard rooms serve as a complex manifestation of broader socio-political disaffection: back-yard tenants are frequently the subjects of complaints to the local police, seen as responsible for increased pressure on infrastructure, burst sewage pipes, and perceived culpability for petty crime. Despite their dependence on back-yard tenants, many Eldorado Park homeowners regard back-yard dwellers as interlopers. At the same time, the precarious and unregulated relationships that emerge between tenants and landlords frequently give rise to abuse. (Calls to the police commonly involve some form of landlord–tenant conflict.)

In its most benign form, the densification of Eldorado Park makes for greater diversity. The racially and culturally homogeneous township is nowadays home not only to black South Africans who have moved into the former 'coloured' township over the past two decades, but also to immigrant communities mostly from Asia and other parts of Africa, many of whom are back-yard tenants. But township residents have had to share (albeit inadequate) space and resources that were previously designated for the use of a very specific group. Schools and clinics are the most obvious and prominent of such resources, although space itself has also had to be re-apportioned. This has been exacerbated by the development of a new low-cost housing project on the border between Eldorado Park and the neighbouring township of Kliptown. Residents from both townships have been promised access to the houses when they are completed. The state has mentioned a figure of 60,000 new homes on land that will be released through various state and municipal departments. As of late 2018, less than one hundred homes stood in various stages of completion, and the development stalled. In May 2018, residents of Kliptown occupied some of the incomplete buildings, drawing a response from the police as well as from residents of Eldorado Park. These compressive stresses are exacerbated by a perceived lack of state or municipal intervention in the township, which taps into more historical perceptions of abandonment among Eldorado Park residents; this is the subject of the following section.

Abandonment

On 3 August 2017, protesters in Eldorado Park set a bus on fire as residents and police engaged in running street battles. This particular protest emerged in response to accusations of racism at a local high school; in particular, the appointment of a black head teacher in what residents claim should be a school for 'coloureds'. Residents cited allegations of racism directed at 'coloured' pupils by black teachers, and a perception that 'coloured' teachers were prevented by the teachers' union from accessing teaching posts in Eldorado

Park (De Villiers 2017). The veracity of the claims and counter-claims are not important for this discussion. What is important is the profound sense of encroachment that residents feel from supposed outsiders to the community. It is all the more poignant that the basis of the claims of inclusion and exclusion are precisely the racial categories that were designed to segregate society under apartheid.

Despite the transition to a multi-racial society in 1994, it is primarily in wealthy parts of the city that racial diversification has been most celebrated (Christopher 2005; Seekings 2010). However, in a context of scarce social resources, race has continued to be the basis for conflict and contestation in many poor parts of the city. Some researchers have suggested that the continued 'ghettoisation' of parts of the city as racially homogeneous and pervasively poor might suggest that the post-apartheid city should perhaps be viewed as a neo-apartheid city (Christopher 2001, 2005). Certainly, the persistence of narratives of racial identification in Eldorado Park means that a majority of Eldorado Park residents with whom we spoke during the research process self-identify as 'coloured' South Africans.

'Coloured' identity in South Africa has an extremely complex relationship to place and language (Adhikari 2005; Goldin 1987; Nilsson 2016; Wicomb 1998). Gillespie (2007) has shown, for example, how 'coloured' identity in the Western Cape is deeply co-implicated with the apartheid-carcereal geography of the failed development of the town of Atlantis, where 'coloured' people were supposed to be relocated/abandoned. Eldorado Park serves as a landscape of abandonment that has inscribed great violence onto the community (and continues to do so, and yet the community remains co-dependent upon it). While 'coloured' South Africans are recognised by the state as black for the purposes of official affirmative action policies etc., in the complex racial geographies of post-apartheid South Africa many residents of Eldorado Park feel neither 'black enough nor white enough' (an expression that we heard on several occasions). In this regard, the scarcity and competition for social infrastructure produced by compressive forces in Eldorado Park, while nominally just a version of the same forces in countless townships and poor urban spaces across South Africa, are also inflicted with a deep sense of despair. In other words, there is a pervasive sense that this is all things will ever be for this community.

Povinelli (2011) conceptualises abandonment not as ignoring or failing to engage, but as an active and deliberate disregard. The two coupled concepts of endurance and exhaustion are crucial to the political economies of abandonment: essentially, it is never simply a story of the weak awaiting the benevolent powerful. Abandonment tests endurance (the possibility of living durable lives, let alone surviving) and exhaustion (the recognition that endurance is usually fatal). The question that the concept of abandonment poses is the possibility of a durable and meaningful life in the context of hegemonic power and the abandonment of bio-political expectations under late liberalism. In this context, residents have attempted a futile re-imagining of the hegemonic state, e.g. via the police station (see the section 'Surveillance' above). At the same time, residents have laid claim to those local state resources that exist, such as schools and hospitals, projecting onto them not a post-apartheid non-racial access, but a strong sense of ownership. Residents do not want to share the meagre services that were apportioned them under apartheid in the post-apartheid context – such is the intensity of feelings of abandonment.

We have seen residents denying access to local clinics and schools to black residents from neighbouring Freedom Park; for some, the use of a clinic designated for 'coloureds' in a 'coloured' neighbourhood should not be one more thing that is taken away from them. We have seen ongoing incidents of residents restricting black educators from teaching their children, accusing them not of incompetence but racism and preferential advantage on the part of government officials because they are black. There is a sense of ownership over their community, being 'coloured' and living in Eldorado Park resonates with their identity, their history of dispossession and institutionalised exclusion under white supremacy continues to shape their interaction with their lived space.

Conclusion

In this chapter, we have presented three viewpoints on Eldorado Park, namely: (1) as a socio-spatial configuration within which people's life-worlds are located, what we have termed 'surveillance'; (2) as a socio-spatial scale at which local residents feel a shared experience, what we have called 'compression'; and (3) as a socio-spatial configuration that reinforces a place-based (and sometimes exclusionary) identity, what we have termed 'abandonment'. In some sense, all that we have done is to present three descriptions of Eldorado Park; three different aspects that are part of the everyday life of this place. However, taken together, we would like to suggest that these three elements constitute the conditions for a particular sub-urban socio-spatial configuration called neighbourhood: that is, a socio-spatial configuration that can contain meaningful life-worlds for its inhabitants, that provides for a largely shared experience of living together, and that reinforces (sometimes negatively, but sometimes in more cosmopolitan ways) socio-cultural identity. Thinking of Eldorado Park in this way does not change what we see – it does not make the deep socio-economic challenges disappear and turn the township into a cohesive and cosmopolitan neighbourhood. Nevertheless, it has allowed us to see these obdurate socio-economic challenges in subtly different ways. Rather than encountering Eldorado Park as an already constituted socio-spatial configuration that fits into a South African urban scholarship norm, thinking aboutof Eldorado Park through neighbourhood allows for the foregrounding of a different set of theoretical and explanatory concepts.

There are three key lessons for us from this exercise:

1. For one thing, looking at Eldorado Park as a neighbourhood (i.e. as a socio-spatial configuration with a defined heart, boundary, and social

identity) shifts one's analysis of it from its relationship to an urban structure to the internal dynamics and relationships that define that socio-spatial configuration. Township implies a relationship to a very particular urban system, and even though the rationality of that system no longer pertains, the effects thereof continue to be felt. In this sense, one might not want to give up the term 'township' because it acknowledges the obduracy of urban forms of inequality (see e.g. Chari 2013). But thinking of Eldorado Park as a neighbourhood allows us to think beyond this structural relationship, without ignoring the ways in which history has shaped this space. Eldorado Park as neighbourhood is a dynamic and changing set of socio-spatial dynamics and relations, even as Eldorado Park as a township is a geography of disconnection and structural dependence.

2. Thinking of Eldorado Park in terms of neighbourhood allows us to think about the similarities between Eldorado Park and other similar urban spaces across the world. This does not imply that any comparison will always be symmetrical, and indeed we are a long way off from any meaningful comparative gesture in our own research. But it is in our view vitally important to shift away from an urban studies tradition in South Africa that focuses on the exceptionalism of South African cities. There are many things that make South African cities peculiar, but there are also many things to be learned from thinking comparatively across contexts (and not just between two different townships, for example). Thinking about Eldorado Park as a neighbourhood opens up the possibilities of ofconsidering Eldorado Park in relation to neighbourhoods in a world of cities. We believe that such a research imperative should be part of the ethics of urban research. That is not to say that all urban research should be comparative: this is by no means the case. But it is to maintain an ethic towards an extroverted sense of place (Massey 1991).

3. Finally, we feel that thinking of Eldorado Park through neighbourhood has allowed us to think differently about the subjects who occupy this space. Whereas the residents of Eldorado Park township might be trapped in structural relationships of exclusion and abandonment, the residents of Eldorado Park neighbourhood are a more complex and agentic community that includes older 'coloured' residents but also a whole range of newer subjectivities – the existence of whom may even serve to break down the boundaries of the township.

None of this is to say that we should dispense with the concept of township when analysing urban space. One cannot understand South African cities without the lens of township. Nor is this to say that we will advocate for the use of 'neighbourhood' as a theoretically meaningful term that requires revising South African urban scholarship. Instead, reflecting on the concept of neighbourhood has forced a critical reflection on the socio-spatial categories that present themselves as self-evident and familiar to us as urban scholars. Eldorado Park is a township; or perhaps more accurately, it is a former township now in a state of becoming. One possible form of this becoming may be as a neighbourhood, or even several neighbourhoods. Whatever it becomes, it is our ethical duty as urban scholars not only to liberate our own mental schema, but to also liberate the individuals who inhabit our cities from the socio-spatial confines that we as scholars subject them to.

References

Adhikari, Mohamed. 2005. *Not white enough, not black enough: Racial identity in the South African coloured community.* Columbus, OH: Ohio University Press.

Aptekar, Sofya. 2017. 'Looking forward, looking back: Collective memory and neighborhood identity in two urban parks.' *Symbolic Interaction*, 40 (1): 101–121.

Augé, Marc. 2008. *Non-place*. London: Verso.

Ballard, Richard, Adam Habib, Imraan Valodia, and Elke Zuern. 2005. 'Globalization, marginalization and contemporary social movements in South Africa.' *African Affairs*, 104 (417): 615–634.

Ballard, Richard and Margot Rubin. 2017. 'A "Marshall Plan" for human settlements: How megaprojects became South Africa's housing policy.' *Transformation: Critical Perspectives on Southern Africa*, 95 (1): 1–31.

Bank, Leslie John. 2011. *Home spaces, street styles: Contesting power and identity in a South African city.* London: Pluto Press.

Benit-Gbaffou, Claire, Laurent Fourchard, and Alex Wafer. 2012. 'Local politics and the circulation of community security initiatives in Johannesburg.' *International Journal of Urban and Regional Research*, 36 (5): 936–957.

Brown, Jennifer. 2013. *Attitudes and experiences of teachers and students towards life orientation: A case study of a state-funded school in Eldorado Park, South Johannesburg.* PhD dissertation. Johannesburg: University of the Witwatersrand.

Chari, Sharad. 2013. 'Detritus in Durban: Polluted environs and the biopolitics of refusal.' In *Imperial debris: On ruins and ruination*, edited by Ann Laura Stoler, 131–161. Durham, NC: Duke University Press.

Charlton, Sarah. 2013. 'State ambitions and peoples' practices: An exploration of RDP housing in Johannesburg.' PhD dissertation. University of Sheffield.

Chatterjee, Partha. 2011. *Lineages of political society: Studies in postcolonial democracy.* New York: Columbia University Press.

Chipkin, Ivor. 2007. *Do South Africans exist? Nationalism, democracy and the identity of 'the people'.* Johannesburg: Wits University Press.

Christopher, Anthony J. 2001. 'Urban segregation in post-apartheid South Africa.' *Urban Studies*, 38 (3): 449–466.

Christopher, A. J. (2005). The slow pace of desegregation in South African cities, 1996-2001. *Urban studies*, 42 (12): 2305–2320.

Cooper-Knock, Sarah-Jane. 2017. 'Beyond Agamben: Sovereignty, policing and "permissive space" in South Africa, and beyond.' *Theoretical Criminology*, 22 (1): 22–41.

Corbridge, Stuart, Glyn Williams, Manoj Srivastava, and René Véron. 2005. *Seeing the state: Governance and governmentality in India.* Cambridge: Cambridge University Press.

De Villiers, James. 2017. 'Racism not core of coloured school protest – Gauteng Education Department.' *News24 online.* www.news24.com/SouthAfrica/News/racism-not-core-of-col oured-school-protest-gauteng-education-department-2017080. Accessed 13 August 2019.

Dirsuweit, Teresa and Alex Wafer. 2005. 'Fear and loathing in Johannesburg: Constructing new urban identities within gated communities.' Territory, enclosure and control: The ecology of urban fragmentation symposium proceedings, CSIR, Pretoria.

Findley, Lisa and Liz Ogbu. 2011. 'South Africa: From township to town.' *Places Journal*, November 2011.

Forrest, Ray, and Ade Kearns. 2001. 'Social cohesion, social capital and the neighbourhood.' *Urban Studies*, 38 (12): 2125–2143.

Gelb, Stephen, ed. 1991. *South Africa's economic crisis*. Cape Town: Zed Books.

Gillespie, Kelly. 2007. *Criminal abstractions and the post-apartheid prison*. PhD dissertation. Chicago, IL: University of Chicago.

Goldin, Ian. 1987. *Making race: The politics and economics of coloured identity in South Africa*. London: Longman.

Government of South Africa. 2004. *'Breaking new ground': A comprehensive plan for the development of sustainable human settlements*. August, 2004. http://abahlali.org/files/Breaking%20new%20ground%20New_Housing_Plan_Cabinet_approved_version.pdf. Accessed 13 August 2019.

Graham, Steve and Simon Marvin. 2002. *Splintering urbanism: Networked infrastructures, technological mobilities and the urban condition*. New York: Routledge.

Haferburg, Christoph. 2013. 'Townships of to-morrow? Cosmo City and inclusive visions for post-apartheid urban futures.' *Habitat International*, 39: 261–268.

Lemanski, Charlotte. 2009. 'Augmented informality: South Africa's backyard dwellings as a by-product of formal housing policies.' *Habitat International*, 33 (4): 472–484.

Mabasa, Hlupheka Michael. 2012. 'Community policing and crime prevention: A community assessment from Eldorado Park, Gauteng.' Unpublished M. Tech dissertation. Tshwane University of Technology.

MacAllister, Iain, Ron J. Johnston, Charles J. Pattie, Helena Tunstall, Danny F. L. Dorling, and David J. Rossiter. 2001. 'Class dealignment and the neighbourhood effect: Miller revisited.' *British Journal of Political Science*, 31 (1): 41–59.

Massey, Doreen. 1991. 'A global sense of place.' *Marxism Today*. June 1991, 24–29.

Matjomane, Mamokete, Koech Cheruiyot, and Samy Katumba. 2018. 'GCRO map of the month August 2018.' Gauteng City-Region Observatory. https://gcro.ac.za/outputs/map-of-the-month/mapping-unemployment/. Accessed 13 August 2019.

Meegan, Richard and Allison Mitchell. 2001. '"It's not community round here, it's neighbourhood": Neighbourhood change and cohesion in urban regeneration policies.' *Urban Studies*, 38 (12): 2167–2194.

Misago, Jean Pierre. 2019. 'Political mobilisation as the trigger of xenophobic violence in post-apartheid South Africa.' *International Journal of Conflict and Violence (IJCV)*, 13: 1–10.

Nilsson, Sara. 2016. 'Coloured by race: A study about the making of coloured identities in South Africa.' Master's Thesis. Uppsala University.

Povinelli, Elizabeth. 2011. *Economies of abandonment: Social belonging and endurance in late liberalism*. Durham, NC: Duke University Press.

Rauch, Janine. 2000. 'Police reform and South Africa's transition.' Proceedings of an international conference on crime and policing in transitional societies, University of the Witwatersrand, Johannesburg. 119–126.

Robinson, Jennifer. 1996. *The power of apartheid: State, power, and space in South African cities*. Oxford: Butterworth-Heinemann.

Secor, Anna J. 2007. 'Between longing and despair: State, space, and subjectivity in Turkey.' *Environment and Planning D: Society and Space*, 25 (1): 33–52.

Seekings, Jeremy. 2010. 'Race, class and inequality in the South African city.' Centre for Social Science Research occasional paper. Cape Town: University of Cape Town.

Smith, David M., ed. 2003. *The apartheid city and beyond: Urbanization and social change in South Africa*. London: Routledge.

StatSA. 2011. *South African Census 2011*, published online 30 October 2012. www.statssa.gov.za. Accessed 24 September 2019.

Swilling, Mark and Eve Annecke. 2006. 'Building sustainable neighbourhoods in South Africa: Learning from the Lynedoch case.' *Environment and Urbanization*, 18 (2): 315–332.

Wafer, Alex. 2012. 'Discourses of infrastructure and citizenship in post-apartheid Soweto.' *Urban Forum*, 23 (2): 233–243.

Wicomb, Zoë. 1998. 'Shame and identity: The case of the coloured in South Africa.' In *Writing South Africa: Literature, apartheid, and democracy, 1970–1995*, edited by Derek Attridge and Rosemary Jolly, 91–107. Cambridge: Cambridge University Press.

Wilhelm-Solomon, Matthew. 2017. 'The ruinous vitalism of the urban form: Ontological orientations in inner-city Johannesburg.' *Critical African Studies*, 9 (2): 174–191.

Wolpe, Harold. 1972. 'Capitalism and cheap labour-power in South Africa: From segregation to apartheid.' *Economy and Society*, 1 (4): 425–456.

2 Killing time in a Roma neighbourhood

Habitus and precariousness in a small town in Western Turkey

Sezai Ozan Zeybek

A place on the side of the road

The photograph shown in Figure 2.1 is of a Roma neighbourhood located on the outskirts of a small provincial town, Malkara, in north-western Turkey. It was taken minutes before midnight, on New Year's Eve. In the photograph, nothing gives that away, though, except for the celebratory writing on the windows: 'Welcome New Year, 2009'. Forget cheering, these men seem gloomy and subdued. On that particular night, they sat around that table, played cards and drank tea until the early hours of the new year, as on any other day. They basically 'killed their time', which was a prevalent activity in this neighbourhood, both when I first went there in 2008 and in my following visits over the years.

The habitual patrons of these coffee shops were men, most of whom had been unemployed for long spells. On an ordinary day, these places would open at 6 am and stay open until midnight. Although most did not have a job to go to, leaving home early and coming here was still a habit of these men. Occasionally they could find jobs in agriculture, on construction sites, or in the open-pit coal mines surrounding the town. They were day labourers. Usually, they would work for a couple of months in a year, and had to wait up to 9 months for the next season. Coal mines were the most reliable source of income, but even there the work lasted only about 6 months. Working in the mines was an extremely hard and backbreaking job that had caused many people to retire at around 45, unable to work elsewhere for the rest of their lives. Some men were disabled from job-related accidents for which they could rarely receive compensation, and ended up in coffee shops for good. In this neighbourhood, men had abundant amounts of time.

This chapter offers an analysis of the temporal formations of a neighbourhood along the lines of class and ethnicity, by focusing on men. For that, I draw on the ethnographic data from one of the Roma neighbourhoods in Malkara, the Hacıevhat neighbourhood. I build my analysis on the observations and stories of young men who have had to wait in different ways. In that regard, I investigate time as a site of subjectification, as well as of alienation, where protagonists are usually waiting for something. They are waiting for

Figure 2.1 Coffeehouse on New Year's Eve.
Source: Photo by Sezai Ozan Zeybek.

someone to call, for a document to arrive, for an opportunity to exploit, and mostly, for a chance to leave town. In all instances, though, waiting denotes 'passing time by doing nothing' for prolonged periods, even years. In this context, I illustrate and discuss the intricacies of a habitus put 'on hold' and barely accruing capital in any form.

The larger region where Malkara is located is one of the most affluent parts of Turkey. The axis from Edirne to Istanbul has succeeded in attracting foreign investment, creating new jobs, and enhancing tax revenues and capital accumulation in the last couple of decades. But Malkara, positioned a little to the south of this axis, is excluded from these dynamic trade routes, production zones, and financial flows. The town lingers as 'a space on the side of the road' (Stewart 1996). It is an 'ordinary' town which has neither importance in, nor relevance to, national agendas. Instead, the town is often rendered 'equivalent' to other towns that exemplify a local variation of a familiar story: underdevelopment, provincial mentality, boredom, etc. This is what Meaghan Morris calls 'the project of positive unoriginality' (Morris quoted in Chakrabarty 2000, p. 39).

The town, with its 25,000 residents, is divided into four neighbourhoods, two of which having a non-nomadic Roma majority population. Although Roma make up one quarter of the town's modern inhabitants, their neighbourhoods are located at the two ends of this peripheral town, to which Turkish origin residents prefer not to enter, if they do not have to. They are marginalised neighbourhoods. Poor and rich Turks alike do not eat Roma food, let flats to Roma families in Turkish districts, send their children to the same schools, or visit coffee shops run by Roma, all of which help to maintain the invisible borders within the town. Roma have not been hired in restaurants to this day. Turkish barbers do not cut the hair of Roma customers, and some restaurants even refuse to serve them food, because using the same cutlery is considered 'repulsive' by many Turks. The word 'Roma' itself is marked (Tannen 2010) and used carefully in public, especially in the presence of strangers. It is often replaced by terms such as 'dark-skinned citizens [*esmer vatandaş* in Turkish]', people from 'the neighbourhood', 'our citizens', 'you-know-who', etc. as a token of the shameful and uneasy relations with them.

In Turkey, Roma are not legally recognised as a minority or a community. Their situation is similar to that of the Kurds, Georgians, Arabs, and Circassians. There are no laws concerning or mentioning Roma, mainly because the republican aspiration in Turkey is a unitary and homogeneous Turkish nation-state (Vali 1996; Sirman 2001; Keyder 2005; Oran 2005; Yeğen 2006). The only legally recognised minorities are Armenians, Rums (Greek-speaking Christians from Anatolia), and Jews, who have all had to live under relentless surveillance, threats, and systematic repression. Being an ethnic minority in Turkey has its long, burdensome history. Therefore, unlike officially recognised non-Muslim minorities or Kurds (with an ongoing political-armed struggle against the Turkish state), Roma in Malkara have been trying to remain under the radar with no claim to a separate ethnic identity. They claimed themselves to be 'patriotic Turks' – though with a Roma background.

Other than systematic, uninhibited discrimination against Roma in Malkara, there is another layer of marginalisation, which is rather a consequence of geographical hierarchies. One of the most frequent questions I received during my 10-month stay there was about why I came to this particular neighbourhood – on the assumption that research should be about somewhere more interesting. 'Nothing happens here', my informants often told me, 'there is nothing to tell about here!' – their eyes probing me to seek out my 'hidden agenda'. The stories that were circulated in newspapers, the events that made a difference, and the people worth talking about were almost always located elsewhere. Living here, according to many of the inhabitants, was too ordinary, too mundane, and too repetitive. They were merely passing time. As if time – as well as space – was split, where some places appeared as agents and makers of history, and meanwhile, other places lost their relevance and importance. They became places full of 'nothing', where time passes without progress or impact. In that regard, 'falling behind' another time and 'living in the shadow' of another place were very much influencing life stories, especially among those with limited resources and symbolic capital. As if the feeling of being 'removed from the spaces where things happen' (Jeffrey 2010, p. 471) coalesced into irrelevance.

These sentiments not only reproduced a hierarchical ordering of people's experiences across space (between centres and peripheries), but also created a dissonance within their life stories, as the present cannot be filled meaningfully to yield a better future. Between a past stained with discrimination and a future that one has little hope about, waiting emerged as the defining attribute of the present, which frequently amounted to doing nothing, wasting time, and loitering in the Roma neighbourhood of Malkara. In other words, it deprived life stories of a meaningful, active present. Their personal narratives were bound to failure, defeat, and submission, not only from the point of view of hypothetical achievers, but also themselves.

Below, I follow two young men and their friends in their daily activities and elucidate different modalities and politics of waiting. Some people are stood up by others and made to wait (Auyero 2012). Some are just waiting for certain adversities to pass. Some others are waiting for an opportunity to exploit, or for some news to arrive (Hage 2009). Yet, whichever modality waiting takes, it transpires under highly asymmetrical power relations.

Killing the present time

Ufuk, a 22-year-old Roma, was married to a young woman from Austria. But after the ceremony in Malkara two years ago, Ufuk and the bride could not see each other again. She was living with her parents in Vienna, working as a dentist's assistant, while Ufuk was waiting for a visa.

The bride's family was originally from Malkara, but had migrated to Austria a long time ago. Her parents wanted Gül, the bride, to marry someone from their hometown in order to maintain ties with their origins, and to preclude a marriage to a 'non-Muslim'. Ufuk was a good candidate. The families had known each other, and Ufuk was one of the few Roma in town with a university degree. So, the marriage was arranged by the families. Ufuk and his wife had not known each other. They had talked on the phone a few times since then, but only briefly. Intimacy was postponed. One time, Ufuk recollected an awkward moment in which he called his wife at home, but he could not be sure if he was speaking with his wife or his mother-in-law. He could not distinguish their voices. He had to say things that could be appropriate for either.

Marrying Gül was a good opportunity for Ufuk, as it gave him a chance to escape the town and live in Europe. Ufuk had been waiting for an EU visa for almost two years when I first met him. He was unemployed, but was not looking for jobs either. His family was able to support him financially during his 'prolonged gap years', perhaps expecting to receive remittances in the long run. He would half-jokingly tell me he would 'hop on a plane and leave the country the day after getting the visa'. He was very keen on the idea of living in Europe because he believed there was 'nothing left in this place'. Jobs were scarce, and his degree made the situation even worse, because Roma were not hired for white-collar jobs in this town.

One day, I walked downtown with five young Roma men, including Ufuk, to do some shopping together. Some shop owners knew me as a 'researcher from abroad', and they were astonished when they saw me 'hanging with Roma'. One middle-aged shopkeeper whispered in my ear that I had to be careful about lice if I intended to keep them by my side. In the bakery, the cashier, after noticing my 'Roma mates', checked if my money was counterfeit, a gesture of unconcealed distrust. A while later, on the very same day, another shopkeeper refused to name the price of a sweatshirt, assuming we were only window-shopping and wasting his time, or worse, disturbing other customers. Two young high school girls passed us on the street, making a conspicuous effort to keep their distance from us, to avoid touching us, as if we were matter 'out of place' (Douglas 2003). Lastly, on our way back to the Roma neighbourhood, we overheard a mother threatening her child that she would give her away to the Roma if she did not behave. She, the mother, was brazen, as if we did not speak their language or as if we were invisible to her. We entered Hacıevhat, the Roma neighbourhood, in sheer silence. This place was also their shelter, despite all its shortcomings. Until midnight, worked up and angry, these five young men told me how they hated to go downtown and how life in Malkara could become unbearable at times. They expressed their wish to leave town and to go someplace where people did not know anything about Roma – a place with no prejudice, no judgement. But unlike Ufuk, none of them had the means or opportunity to do so. Ufuk consoled himself by saying he was at least waiting for a temporary period, because for others, waiting could easily turn into waiting in vain.

Occasionally, Ufuk would do some manual labour for pocket money. However, I soon realised that working was not really the opposite of waiting. Instead, both working and waiting were organised around an abundance of time. The value of the labour, and the value of those who laboured, were deflated by the systematic utilisation of their abundant time. Abundant time signified powerlessness: the more time there was, the more it could be devalued and exploited. As a result, particular lives, with scarce resources but a lot of time, are kept in a perpetual position of waiting even in work relations.

One chilly day in November, one of the big landowners of the town contacted Ufuk and asked for four people to transfer 15 tonnes of animal feed from a truck to a depot. I decided to join the team. The landowner said he would pick us up at 9 in the morning. The next day at 8:30 am we met up to wait for the employer. The rest of the workers (Ufuk, Tuncay, and Emrah) had their worn-out work clothes on, suited to 'dirty' jobs, unlike my everyday clothes. After waiting until 10 am, Ufuk finally called the landowner. He reassured Ufuk and said that he would eventually come, but that there were some problems with the truck. The conversation was very short: the employer at the other end did most of the talking and Ufuk just nodded and said that we were waiting.

Around noon, patrons of the coffee shop started to make jokes about us – that our waiting was in vain and that we had better change our clothes. We had all become a bit bitter and decided to ask for more money as compensation. Yet, no one among us in fact knew how much we were meant to be paid in the first place. The negotiation was going to take place only after the job was done. Therefore, ambiguity around the question of 'when' compounded the question of 'how much'.

By the afternoon, the employer still had not shown up, so we decided to go downtown and check with him at his office. When we found him, he told us that the job was delayed and that we had to wait until tomorrow. He had probably known this since the morning, but did not bother to call us. He was dismissive all along. I was pissed off when we left his office. But others were happy to finally be able to change out of their dirty overalls at 4 pm. We spent the rest of the evening together in the coffee shop playing more card games and drinking more tea.

The next day, we met at 9 am at the same spot. We learned that Emrah and Tuncay had gone to the coal mine that day. Ufuk replaced them with Rauf and Hakan, who were 'roaming around'. But the employer did not turn up again, so we had to wait until noon without a definite time plan.

For me, it was very hard to wait for such long hours. Time passed in the slowest way possible. Playing cards and chatting could only kill so much time, especially when one is expecting something. I could not help but push Ufuk to call him again. The employer finally informed us that the truck had arrived at the depot in a nearby village. We arrived there at around 1 pm. We had to wait for another 45 minutes for the landowner to give us the go-ahead.

The job was hard and exhausting, at least for me. I was obviously the weakest link. In 3 hours, we unloaded the truck and carried everything to the depot. By the end, I could barely breathe. My co-workers were smiling at me and praising my effort, but I knew I had been nothing but a burden all along. The negotiation took place at the end. The landowner asked us to value our own work and to name a price. He added: 'We will always be in need of each other. Don't name a high price, or I won't call you again.' Ufuk asked for 50 Lira (≈US$9) for each of us. The employer mentioned the economic crisis and how hard it was to earn a living these days as part of the negotiation. In the end, we received 40 Lira (≈US$7) each. Before we left, he gave us an extra 5 Lira (≈US$1) as a gesture: 'Take this, too. You are now indebted to me for the next time', he told us. We took the money, thanked him, and left.

With or without work, Ufuk was waiting most of the time. In this regard, Ufuk's time was not merely 'free time', which is often conceptualised as an integral part of capitalist temporality, but rather, it was 'dead time'. In the conventional stories, while 'shrinking time' fills the metropolitan centres (for prominent examples see Harvey 1988; Castells 1996), provincial places and dispossessed lives are filled with 'free time' that is subsumed under the time of production. Conversely, Ufuk's time was neither shrinking nor was it free. Ufuk's present time was rather dead, because time was something to kill. According to Pierre Bourdieu, dead time signifies a specific temporal void in the social structure. He describes dead time as a purposeless and meaningless time that leaves people with despair and anxiety:

The empty time that has to be 'killed' is opposed to the full (or well-filled) time of the 'busy' person who, as we say, does not notice time passing – whereas, paradoxically, powerlessness, which breaks the relation of immersion in the imminent, makes one conscious of the passage of time as when waiting.

(Bourdieu 2000a, p. 224)

Powerlessness in the form of waiting was felt in other realms as well, especially in bureaucracy: long waiting times were emblematic of the treatment Roma people received. It would usually take a whole day to pay a bill, to deliver a document, to receive a benefit (as seen in Figure 2.2), or to get medicine. For the most part, they would have to wait in front of an office, sometimes for hours. Queues were long; problems were complicated. In the meantime, it was hard to be rid of the feeling of inferiority. Loathing towards Roma cut across class, gender, age, and education among Turkish-origin people, and among officials as well, where it mattered even more. My Roma acquaintances frequently asked me to come along when they had a bureaucratic problem to solve, so as to empower themselves, and if possible, to speed up the process. I was, at the end of the day, a 'Gaco' – a respected Turk.

Figure 2.2 A day at the social services, waiting for the coal aid.
Source: Photo by Sezai Ozan Zeybek.

Lack of jobs, lack of money, and other forms of material deprivation obviously restricted the possibilities for a full present and generated a sentiment of 'redundancy'. Yet, material deprivation in itself cannot explain the full extent of this sentiment. For one, not all people shared in it to the same degree, nor could everyone kill time. Women, for example, living in the same neighbourhood, led busier lives. They would do the shopping, cooking, cleaning, and caring throughout the year. They were not 'out of a job'. Although poverty was common, their complaints were different as well, ranging from domestic violence to children's education and even to duplicitous relatives that they had to accommodate. Indeed, in a different light, Roma men, too, were doing a lot during the day: looking for jobs, networking, asking for favours, returning favours, repairing things, navigating bureaucracy, and facing all the regular, arduous problems of everyday life. In that regard, the present was not an existential void (see Goodstein 2005, to see its difference from ennui), but a function of a particular spatial hierarchy that almost always attributes failure and superfluity to peripheries. Put differently, although large chunks of dead time really had an impact on the texture of the neighbourhood, there was also a larger context that contributed to the sentiment of 'living a life not worth telling', or 'being left-behind at the periphery'. The ongoing comparisons between more glamourous centres, where supposedly most things happen, and Malkara, where supposedly nothing happens, derive from a particular set of significations with regard to the value of time of respective geographies. Let me expand on this point.

The invasion of timeless times

One of the common traits of a diverse range of hegemonic narratives, such as globalisation, capitalism, modernity, or westernisation, is that they all signify a movement, or rather, 'penetration' of an overriding set of relations to other places, which one might call diffusionism (Blaut 1993, 2000a). They 'unfold' in different places with an inherent logic, by which previous (and usually trivial) traits of peripheral places are replaced with what capitalism, westernisation, globalisation, etc. bring forth. In such discourses of unfolding, the degree of 'diffusion' is assumed to be the primary mover of history; that is, history becomes the history of the incorporation of other places. Peripheral places are ranked according to their levels of development, globalisation, or modernity. Difference becomes a modality of movement, as the majority of places are at the receiving end: they only make up the echoes of an original moment in history that already took place somewhere else, be it modernity or globalisation. They may be *added to* the main story as examples, but cannot be *added up* to something novel. They merely become more of the same.

Gibson-Graham argues that such encompassing historical scripts, which are used as depositories for explaining historical progress, make it a lot more difficult to think of anything beyond these scripts (Gibson-Graham 2006). They appear to be everywhere, or about to arrive, and they fill in 'empty' spaces with their own logics of expansion. Even critical accounts, such as Marxist accounts, share a common vocabulary with what they oppose: an ever-expanding capitalism.

> The standardized and dominant globalization script constitutes … sites that may be recalcitrant but are incapable of retaliation, sites in which cooperation in the act of rape is called for and ultimately obtained … To accept this script as a reality is to severely circumscribe the sorts of defensive and offensive actions.
>
> (Gibson-Graham 2006, p. 126)

Gibson-Graham notes the phallic connotations of this type of language, which frequently refers to invasion, penetration, subordination, and pre-constituted potential victims. This language, according to them, suggests an inevitable rape. The position of the so-called victims (peripheral countries, women, pre-modern places, provincial towns, and so forth) becomes secondary, because the true act of the scenario lies with the male (often white) protagonists. The invasion is almost inevitable, and even necessary. Remember the Communist Manifesto's famous expectation that capital melts all the solid barriers that have stood in its way with its relentless and tireless dynamics; expands its reach to all the material world we are familiar with by subjecting all realms under its power (Marx 1848; see also Coronil 2000 for a discussion).

Or read Marshall Berman's observations, for example, this time about modernity: 'Modern environments and experiences cut across all boundaries of geography and ethnicity, of class and nationality, of religion and ideology. In this sense, modernity can be said to unite all mankind' (Berman 2000, p. 15). These accounts see the 'invasive' powers as the condition of revolution and change (which may bring forth negative or positive outcomes, depending on the view of the interlocutor), yet, inevitably, they 'melt away' archaic relations, traditions, local ties, 'the old'.

These narrative conventions lend themselves easily to the old catch-up game, where provincial places are expected to 'catch up' with modernity, globalisation, westernisation, and so forth and, until that is accomplished, they must wait in anticipation and bear the brunt of the burden of 'being behind'. Yet, 'catching up' rarely fulfils its promise, because the whole notion of catching up is based on a strictly preserved distance composed of temporal and spatial ruptures, which is not to overcome but to preserve (Ahıska 2003). While some may enjoy particular privileges of dominant spatial-temporal formations, others are systematically kept outside. As a consequence, most of the world (developing countries, modernising societies, etc.) is kept in an enduring waiting room of provinciality. These societies become incarcerated in a never-ending process of catching up with those times that are always running ahead. All along, these narratives overwrite the diversity of potential stories, minor practices, and tactics. Diffusionist narratives become 'the essential but usually unquestioned background to our stories' (Massey 1999, p. 34).

With all [the] emphasis on flux and flow, this literature tends to blind itself … as they render certain places incontestable centers, and others, undeniable peripheries. It glosses over the simple social fact that for some, a sense of 'incarceration' precedes ethnographication.

(Blank 2004, p. 356)

To return to our initial discussion: the dominant spatial-temporal representations divide the world into 'makers of history' and 'receivers of history' and this is also why particular places are perceived to be lagging behind other times and other places. 'Peripheries', in Dipesh Chakrabarty's words, are perceived as occupying an imaginary 'waiting room of history' (Chakrabarty 2000, p. 8). In short, the present of a small provincial town and its most marginalised neighbourhood becomes worthless and dispensable because it exists in anticipation of other times – especially if one's eyes are fixed on these other places.

In the next section, I will recount another story, this time of Rauf, who was seduced by these other places, as so many other young people in Malkara. But being from the Hacıevhat neighbourhood made his exit all the more difficult, because his limited means did not meet his aspirations and kept him in a waiting room. He refrained from participating in neighbourly networks, but could not break free, either. That, I argue, also broke his 'temporal pragmata of habitus', to borrow the term from Pierre Bourdieu, meaning that the relation between past experience and expectations of the future came to a point of disjunction (Bourdieu 2000b).

Millennial aspirations and casino capitalism

Rauf, 29 years old, had studied veterinary medicine, but had not practised his profession since graduating. Like Ufuk, he could not find a job in Malkara when he returned because all veterinary jobs were (unofficially) reserved for people of Turkish origin. His ultimate desire was to leave Malkara. He wanted to go and live in a big city, and if possible, in Europe. But this required money, a lot of money that he did not have.

For a year, he had worked on a construction site in Dubai, under the worst possible conditions, by his recollection. Along with the other workers, he had stayed on the construction site the whole time. It was outside of Dubai, and there was no transportation between the two. Therefore, he could not even visit the city. Under the desert sun and against the sand storms, they worked up to 14 hours a day. At least the pay was good. He returned home with a few thousand US dollars in his pocket. However, as he repentantly told me, that money was used up within a few months, mostly on old debts, girls, and expensive consumer items. After a while, he could not even pay his credit card debts. In the end, he got 'stuck in that neighbourhood' with his incensed father.

Later, he had a second opportunity to 'hit big money'. By mobilising his college circle, he found a summer job at a 5-star hotel in Antalya, a tourist destination in Southern Turkey. He officially worked there as a masseur, but as he put it, the job description included 'pleasing older women' as well, who were mostly from wealthy European countries. The hotel did not pay him, but Rauf claimed that he earned tips up to 100 US dollars from his satisfied customers, and that it was like his dream job.

Yet, he had been laid off and come back to Malkara for reasons he would not share. His stories involving women were creating controversy in the neighbourhood. His father cursed every time his name was mentioned, and religious people, in particular, were avoiding him on the street. His ambition was to marry a European woman, and in this way, to obtain a visa. 'Everything is better in Europe', he once told us: 'even sex is better'.

He slept until noon most days, then spent the rest of the day in an Internet café. He would stay there for hours, playing games and chatting with online friends. He also had a few 'online girlfriends', with whom he 'hung out'. But he was not planning to marry despite pressure from his father. He wanted to 'enjoy life while he was young'. At night, he would drink, smoke joints, play card games, and occasionally visit the brothel in the neighbouring town.

One day, Rauf announced his next project. He was planning to go to Russia and work for an Irish construction company. 'Everything was set', he told us in the coffee shop. A friend of a friend promised to sign him up in the next couple of months. Rauf's ultimate aim was to make his way to Europe, and for him, Russia was Europe. But first, he was supposed to send his passport to this man, along with one hundred dollars in cash.

When I first heard about this, I could not help but think that this seemed more like a textbook scam: someone with no name collects money with no papers. I warned Rauf and offered him to call the Irish company to learn if they really had a construction bid in Moscow. He refused. He was sure this was a safe bet. I suggested that he at least learn the name of the company to check if it really existed. He refused that, too. In the end, he lost his passport and his one hundred dollars.

At first, I found his refusal to acknowledge any likelihood of a scam very unreasonable. 'In a world gone awry', as Comaroff and Comaroff argue, it was perhaps typical to witness 'extreme enthusiasm among the jobless youth towards trying their luck to make a fast buck' (Comaroff and Comaroff 2000, p. 297). Indeed, he persistently came up with other 'lucrative projects' every time we met: selling his fully equipped, high-level avatar from an online game he was playing, treasure hunting in the old Armenian district of the town, trading a new Chinese gadget he saw on the Internet with high profit margins before its market thrives, thinking of an idea no one had thought of yet, or investing in an innovative company of one of his friends, which would definitely bring big money in no time. He was drawn to the allure of accruing wealth quickly, similar to the casino capitalism of high finance, but without capital (ibid. pp. 309–314).

I soon realised that he was desperately trying to do something, instead of nothing, but there was a dissonance between what he aspired to and what he could get. As Arjun Appadurai succinctly observes, even aspirations are characteristically a function of social positions, and are not evenly distributed in society. To quote him at length:

> [T]he better off you are (in terms of power, dignity, and material resources), the more likely you are to be conscious of the links between the more and less immediate objects of aspiration. Because the better off, by definition, have a more complex experience of the relation between a wide range of ends and means, because they have a bigger stock of available experiences of the relationship of aspirations and outcomes, because they are in a better position to explore and harvest diverse experiences of exploration and trial, because of their many opportunities to link material goods and immediate opportunities to more general and generic possibilities and options.
>
> (Appadurai 2004, p. 68)

Linking resources to expected outcomes does not come naturally. It is part of a habitus harnessed over the years and accumulated by (intergenerational) experience. It also has a temporal dimension. Bourdieu, in his *Pascalian Meditations*, dwells on the temporality of habitus as the 'presence of the past in the present which makes possible the presence of the forthcoming in the present' (Bourdieu 2000b, p. 210). Put differently, habitus is shaped by our past, but also enables us to act by anticipating the forthcoming in the immediate present. It is the capacity to synchronise past, present, and future to act 'well'. Such a capacity is not necessarily the same as 'planning ahead' because it stems from pragmata – a practical sense of the conditions, as opposed to a rational calculation of the future.

Therein, we can say that the notion of habitus emerges at the nexus of different temporalities. It adjusts expectations to chances (objective conditions), but in doing so, habitus also ensures 'the unconditional submission of the dominated to the established order ...' (Bourdieu 2000b, p. 231). Any dissonance in habitus might result in exclusion from the social field as well as in humiliation and condemnation. But at the same time, the same dissonance might bring forth new possibilities for challenging the established order in new ways, and it might designate new subject positions that contest the status quo.

In this respect, Rauf's story might be interpreted in two ways. In the first interpretation, the discrepancy between his ambitions and his objective conditions causes a generalised and lasting disorganisation and incoherency in his behaviours, which leads to disappointment. Rauf, with his dream-like aspirations (from the point of view of his father) and his 'millenarian hopes', lives in a fantasy world. In fact, some of the projects he entertains were, as Bourdieu would put it, 'completely detached from the present, and immediately belied by it' (Bourdieu 2000b, p. 222).

The second interpretation goes against the conventions and comforts of habitus and disrupts the reproduction of the social order. Rather than pursuing hopeless projects, Rauf was taking a different route to get what he wanted.

Despite all the risks of humiliation, he was not performing 'the patriotic Roma victim', unlike the majority of the neighbourhood. He was not 'making nice' with the big men of the town – who abuse Roma at almost every turn – in anticipation of a favour, which would seem only 'reasonable' given the circumstances. He was not passively accepting the insults that came his way. He was not taking the life he was expected to take. Instead, he was reaching out in anticipation and against all odds, which could leave 'a margin of freedom for political action aimed at reopening the space of possibles [sic]' (Bourdieu 2000b, p. 234).

The point here is neither to romanticise Rauf's attempts nor to disparage the people waiting in coffee shops. The point, rather, is to see Rauf in different shades of light, and in comparison to other stories, without turning him into an emblematic example of resistance or failure. The life he wanted (with girls, cars, and quick money) was perhaps also questionable at various levels, and not so different from 'casino capitalism' (Comaroff and Comaroff 2000). But the 'reasonable' alternative, I want to point out, shrinks the horizon, imprisons him both spatially and temporally in the neighbourhood, plays into the hands of privileged groups through patronage relations and keeps marginalised groups as cheap labour. He would be, then, also kept out of sight. And when there is a rather unavoidable encounter; he was, like the other Roma of the neighbourhood, expected to act as a 'patriot' and/or 'grateful citizen'. Although one can argue that the game cuts both ways, with influential men having to offer favours every now and then (and that is the very politics of the governed – see Chatterjee 2004; or also Zeybek 2012), at the end of the day, Roma are kept in place. Their aspirations diminish while their time swells within the spatial boundaries they cannot transgress.

Conclusion

In this chapter, my main attempt was to illustrate how the 'abundance of time' that builds up in this Roma neighbourhood was an effect of a specific power configuration. Time, as such, was deeply ingrained in discriminatory practices and subject to devaluation. In this light, I looked into the temporal dimension of deprivation and tried to understand quotidian experiences of precariousness in a peripheral neighbourhood. I used 'dead time' (rather than 'free time') as my guiding concept, which can be defined as the passage of time with no apparent prospects of accumulation or a sense of momentum. With that, I tried to differentiate between different modalities of marginalisation – especially of those who are not fully exploited under labour relations, and who, as a result, are bordering upon redundancy (Hoogvelt 1997). I focused on how different forms of waiting stem from the abundance of time, how waiting becomes the prevalent activity of so many people, and how it conveys sentiments of inferiority and diminishes aspirations.

Apparently, even aspirations are distributed disproportionately. Killing time in a coffee shop from dawn to dusk, and for years, makes future prospects less attainable and more incoherent, as Appadurai observes, 'not because of any cognitive deficit on the part of the poor but because the capacity to aspire, like any complex cultural capacity, thrives and survives on practice, repetition, exploration, conjecture, and refutation' (Appadurai 2004, p. 69). Thus, aspirations have a strong class dimension.

I also tried to show how at times discrepancies between ambitions and means, or rather between past, present, and the forthcoming, result in the loss of pragmata, 'the feel of the game', which sever ties to so-called reality, and further decrease the life chances of deprived people. Severing ties also has the potential to disrupt the established order, change the rules of the game, and interrupt the reproduction of social hierarchies. But without a specific political agenda and coordinated effort, this seems less likely to happen. Precariousness in provinciality can easily go astray in quick-buck schemes, self-depreciation, and violence or hatred against other deprived groups.

A second dimension of provinciality I have sought to problematise attends to larger geographical representations and diffusionist narratives. I argued that these make up another stratum of devaluation, next to material poverty, that depicts the majority of the world as trivial, transitory, and lagging behind. As Johannes Fabian compellingly showed, relations between different places are often also posed as temporal relations, which disavows co-evalness (Fabian 1983). These representations also reverberate in the peripheries. Coveting other lives and feeling stuck become the pathos of the place, especially for those who still desire to leave it. But this pathos, in return, can hollow out the present time, here and now, and reduce these sites to a waiting room, especially for those with fewer resources.

All along, my starting proposition was that agents do not just live in time but make time in a particular spatial configuration. However, as in any other social construct, 'making time' does not mean individuals have complete autonomy in relation to it. There are institutions involved; histories, partitions, arrangements that accelerate or slow it down. Even the succession of time – the past, present, and immediate future, and how these are connected to each other – follows the contours of class differences and geographical hierarchies. Its valuation depends on whether and how one fills it or where one is located. In this regard, by studying different modalities of waiting in this Roma neighbourhood, I illustrated the temporal aspects of the formation of a Roma neighbourhood and its hierarchies.

Coda

> *Four young Roma men and I are sitting around a table. We are chatting. At one point, Selahattin takes out a candy bar, opens it slowly, takes one bite, and passes it to his friend next to him. That meant that everyone would have a bite. While we're talking about this and that, the candy bar is passed around. It's my turn. Selahattin is sitting right across the table, he keeps on talking but his eyes are fixed on me. He is waiting for my next move. When*

I'm biting into the candy bar, he stops talking, as if he can't concentrate on what he's saying while he's watching me eat. I pass the candy bar on to the next person, but Selahattin intercepts. He wants to rewind that particular moment. 'Give it back to Ozan, come on give it back, he should have another bite, he really liked the candy bar', he says. It comes back to me. I take another bite. Time freezes. You can hear a pin drop around the table. Selahattin is watching me, with a great big smile, pleasure radiating from his face. For a short moment, we occupy the same space and the same moment.

References

Ahıska, Meltem. 2003. 'Occidentalism: The Historical Fantasy of the Modern.' *The South Atlantic Quarterly* 2/3 (102): 351–379.

Appadurai, Arjun. 2004. 'The Capacity to Aspire: Culture and the Terms of Recognition.' In *Culture and Public Action*, edited by Vijayendra Rao and Michael Walton, 59–84. Stanford, CA: Stanford University Press.

Auyero, Javier. 2012. *Patients of the State: The Politics of Waiting in Argentina*. Durham: Duke University Press Books.

Berman, Marshall. 2000. *All That Is Solid Melts into Air: The Experience of Modernity*. London: Verso.

Blank, Diana R. 2004. 'Fairytale Cynicism in the "Kingdom of Plastic Bags": The Powerlessness of Place in a Ukranian Border Town.' *Ethnography* 5 (3): 349–378.

Blaut, James Morris. 1993. *The Colonizer's Model of the World: Geographical Diffusionism and Eurocentric History*. New York: Guilford Press.

———. 2000a. *Eight Eurocenric Historians*. New York and London: The Guilford Press.

Bourdieu, Pierre. 2000a. *Pascalian Meditations*. Cambridge: Polity.

———. 2000b. 'Social Being, Time and the Sense of Existence.' In *Pascalian Meditations*, edited by Pierre Bourdieu and translated by Richard Nice, 206–245. Cambridge: Polity.

Castells, Manuel. 1996. *The Information Age: Economy, Society and Culture Vol. 1 The Rise of the Network Society*. Cambridge, MA: Blackwell Publishers.

Chakrabarty, Dipesh. 2000. *Provincializing Europe: Postcolonial Thought and Historical Difference*. Princeton: Princeton University Press.

Chatterjee, Partha. 2004. 'The Politics of the Governed: Reflections on Popular Politics in Most of the World.' University Seminars/Leonard Hastings Schoff Memorial Lectures. New York: Columbia University Press.

Comaroff, Jean, and John Comaroff. 2000. 'Millennial Capitalism: First Thoughts on a Second Coming.' *Public Culture* 12 (2): 291–343.

Coronil, Fernando. 2000. 'Towards a Critique of Globalcentrism: Speculations on Capitalism's Nature.' *Public Culture* 12 (2): 351–374.

Douglas, Mary. 2003. *Purity and Danger: An Analysis of Concepts of Pollution and Taboo*. Mary Douglas: Collected Works. London: Routledge.

Fabian, Johannes. 1983. *Time and the Other: How Anthropology Makes Its Object*. New York: Columbia University Press.

Gibson-Graham, Julie Katherine. 2006. *The End Of Capitalism (As We Knew It): A Feminist Critique of Political Economy*. Minneapolis: University of Minnesota Press.

Goodstein, Elizabeth S. 2005. *Experience without Qualities: Boredom and Modernity*. Stanford and California: Stanford University Press.

Hage, Ghassan. 2009. *Waiting*. Print on Demand edition. Carlton, Victoria: Melbourne University Publishing.

Harvey, David. 1988. *The Condition of Postmodernity: An Enquiry into the Origins of Cultural Change*. Oxford: Basil Blackwell.

Hoogvelt, Ankie. 1997. *Globalisation and the Postcolonial World: The New Political Economy of Development*. Houndmills, Basingstoke and Hampshire: Macmillan.

Jeffrey, Craig. 2010. 'Timepass: Youth, Class and Time among Unemployed Young Men in India.' *American Ethnologist* 37 (3): 456–481.

Keyder, Çağlar. 2005. 'A History and Geography of Turkish Nationalism.' In *Citizenship and the Nation-State in Greece and Turkey*, edited by Faruk Birtek and Thalia G. Dragonas, 1–17. London: Routledge.

Marx, Karl. 1848. 'The Communist Manifesto.' Mondopolitico. www.mondopolitico.com/library/communistmanifesto/communistmanifesto_intro.htm.

Massey, Doreen. 1999. 'Imagining Globalisation: Power-Geometries of Time-Space.' In *Global Futures: Migration, Environment, and Globalization*, edited by Avtar Brah, Mary J. Hickman and Mairtin Mac an Ghaill, 27–44. London: Palgrave Macmillan.

Oran, Baskın. 2005. *Türkiye'de Azınlıklar: Kavramlar, Teori, Lozan, İç Mevzuat, İçtihat, Uygulama*. Istanbul: İletişim Yayınları.

Sirman, Nükhet. 2001. 'Sosyal Bilimlerde Gelişmecilik ve Köy Çalışmaları.' *Toplum ve Bilim* Spring 88: 251–254.

Stewart, Kathleen. 1996. *A Space on the Side of the Road: Cultural Poetics in an 'Other' America*. Princeton: Princeton University Press.

Tannen, Deborah. 2010. 'Marked: Women in the Workplace.' In *Practical Skeptic*, 5th Revised edition, edited by Lisa J. McIntyre, 131–137. New York and London: Mcgraw-Hill Higher Education.

Vali, Abbas. 1996. 'Nationalism and Kurdish Historical Writing.' *New Perspectives on Turkey* 14: 23–51.

Yeğen, Mesut. 2006. *Müstakbel Türk'ten Sözde Vatandaşa*. İstanbul: İletişim.

Zeybek, Sezai Ozan. 2012. '"Fraudulent" Citizens of a Small Town: Occidentalism in Turkey.' *Antipode* 44 (4): 1551–1568.

3 Of Basti and bazaar

Place-making and women's lives in Nizamuddin, Delhi

Samprati Pani

Introduction: the changing South Asian cityscape

The twenty-first century is believed to be a century of the urban, with more and more of the world's population living in cities and more areas becoming urbanised. India's urban population is expected to reach 590 million by 2030 (as cited in Shaw 2012), and it is speculated that 590,000 square miles of land will be taken over by cities globally in the next two decades (as cited in Banerjee 2016). Delhi, the second-largest urban conglomeration in the world, is characterised by many churnings and contradictions in its spatial transformation. This chapter focuses on one locus of place-making in Delhi's neighbourhoods—the associative milieu of weekly bazaars—to underscore the diverse and dynamic spatialities of the ordinary city.

In the context of urban studies in India, Donner and De Neve (2006, 7) draw attention to how a large body of literature has considered place, more specifically neighbourhoods, as sites or backdrops, while the 'real' objects of study are something else, for instance, caste, community, or migration. They further argue that the renewed interest in spatial transformations in the context of globalisation across disciplines tends 'to treat the local as a less central concern in contemporary social analysis' (Donner and De Neve 2006, 2). The globalisation literature has been valuable in showing how conflations of place, community, culture, and identity are problematic, and its concepts such as placelessness, deterritorialisation, non-places, and instantaneity are productive for understanding global processes. However, as pointed out by Escobar (2001, 141), globalocentrism has the danger of erasing place, which 'has profound consequences for our understanding of culture, knowledge, nature, and economy'. This chapter argues not only that place continues to be important in how the ordinary city is lived, used, and experienced but also that place cannot be considered as given, either as a local backdrop or as already determined by global structures. In doing so, the chapter draws inspiration from and is located within a growing body of literature that, since the 1990s, has drawn attention to the need to understand how places are socially constructed and what relationships they make possible—from the work of feminist geographer Doreen Massey's anti-essentialist conceptualisation of place (Massey 1994) to the spatial turn in the social sciences in general (see, for instance, Warf and Arias 2008) and in anthropology in particular (Gupta and Ferguson 1992, 2001 [1997]; Escobar 2001).

In an era when while residing in an urban village in south Delhi, you can eat your favourite Thai food at the local shopping complex; stay connected with 'friends' and relatives all over the world, some of whom you will never meet, through social media; buy crackers made in China from a roadside vendor to celebrate the festival of Diwali; work as a proofreader for a multinational publishing house with its headquarters in London; and get your printer fixed remotely by a call centre employee based in Bengaluru, does it still make sense to talk of the local? Such examples of our lives being entangled with places elsewhere are oft repeated in scholarly works as well as in everyday conversations as illustrative of the space-time compression in an era of global connectedness. While it has a real basis in how everyday lives are changing, what needs to be recognised is that such an experience is neither inevitable nor universal. Take another instance of a life in Delhi that involves waiting for the weekly bazaar in the locality to buy the week's supply of fresh vegetables, going for walks in the neighbourhood park, chatting with neighbours while dropping the kids at the bus stop, gossiping with colleagues at the chai shop on the street opposite the office, spending Sundays with the family at a picnic on the India Gate lawns. This itinerary does not preclude the former, nor is it being cited here as an example of a more authentic or vernacular mode of being. The two examples underline that there is nothing fixed about the degree and nature of connectedness with or disconnectedness from the local and the global. And much of the everyday routines and practices of people in cities across the world still revolve around places of work, leisure, and domesticity. People make their 'biographies in time and space through the routines of everyday life' (Warf and Arias 2008, 1), making and remaking their social worlds in the process.

This chapter draws on Escobar's conceptualisation of place as 'the experience of a particular location with some measure of groundedness (however, unstable), sense of boundaries (however, permeable), and connection to everyday life, even if its identity is constructed, traversed by power, and never fixed' (Escobar 2001, 140). It is precisely because places are experienced, used, and constructed—materially, associatively, imaginatively, and affectively—by different people in different ways and in different situations that they are significant as objects of study: 'all associations of place, people and culture are social and historical creations to be explained, not given natural facts' (Gupta and Ferguson 2001 [1997], 4). Herein also lies the key to conceptualising and methodologically approaching places as neither fixed sites of an unchanging 'local' nor as already overdetermined by larger global structures, but as continuous *processes of making* involving movements, practices, and encounters between people, objects, and materialities. These processes co-constitute dynamic associative milieus around different loci of place-making. Such a processual approach to places also provides a way to steer clear of the overdetermined categories of Western, South Asian, developed, developing, global, and mega cities and instead think of 'a world of ordinary cities, which are all dynamic and diverse, if conflicted arenas for social and economic life' (Robinson 2006, 1).

The chapter moves between the making of two kinds of places, the weekly bazaar and the neighbourhood—how the two are implicated in each other's making as well as how this making is entangled with other places in the city and beyond. The focus is not on exchanges, practices and relationships located *in* a place per se but on how these are shaped by and, in turn, continuously make a place. The next section examines how weekly bazaars, despite the precarity of street vending, have been an enduring feature of the city of Delhi. It shows that this endurance should, however, not be seen as emblematic of the continuance or resilience of 'traditional' practices but in terms of how it makes certain relationships possible. Using illustrations of various routines that customers develop around weekly bazaars, the third section shows that the co-presence of the weekly bazaar and the neighbourhood is central to these routines and the continuous making of weekly bazaars. The fourth section looks at a specific neighbourhood in Delhi, Nizamuddin Basti, and the different associative milieus of place-making through which the neighbourhood is experienced and constructed, arguing that there is nothing fixed or essential to how a place is known. The final section looks at the associative milieu around the Monday bazaar in the Basti—how the affects and practices of the Basti's women fabricate the bazaar as *theirs* and *of* the Basti, intertwining the making of the neighbourhood and the bazaar.

This chapter is part of a larger research project on the making of weekly bazaars through the different practices of vendors and customers, which fabricate the bazaar as *a place in process*. Fieldwork for the research was conducted across weekly bazaars in different neighbourhoods of Delhi, over a period of 16 months in 2016–18.

Weekly bazaars in Delhi: survival of a past or places in making

Almost every locality in Delhi has makeshift weekly bazaars on a fixed day of the week, offering a motley variety of commodities such as fresh vegetables and fruits, clothes, lingerie, glass bangles, plastic goods, utensils, pickles, spices, cosmetics, and street food and services such as repair of zips, bags, and kitchen appliances. Through these bazaars, a city space with a more or less fixed use on other days of the week is transformed into a bustling marketplace and a space of leisure and pleasure.

Despite the legal ambiguity and precarity of these bazaars, and the continuous negotiations and conflicts of these bazaars with state authorities and civil society associations, the number and size of weekly bazaars has grown over the years. Many of these bazaars have been operating in the same locality for as long as 20–40 years, with the location of various individual vendors within the bazaars remaining unchanged over the years. Anjaria, in the context of spatial conflicts around street vending, argues that it is important to understand 'how people stay in place as much as how they are forced out' (2016, 19).[1] This chapter argues that the veracity of people staying in place also has implications for continued connections with places and the making of places. It is the relative stability and the periodic nature of weekly bazaars that makes it possible for various kinds of relationships, practices, and routines to develop, which in turn shape the character of these bazaars.

Despite, or in addition to, the diverse retail options Delhi offers, weekly bazaars are seen as an iconic part of the city's consumption practices as well as the city itself. One newspaper article claims that 'these markets represent a history, a tradition and a cultural continuity' (Hashmi 2007), while another states that 'Delhi's real shopping happens not in malls but at the weekly street markets' (*Economic Times* 2013). The contrast between 'traditional' and new shopping practices is a recurrent theme in the reportage of weekly bazaars. For instance, an article titled 'Delhi's Swanky Shopping Malls No Match for Weekly Bazaars', which refers to an Associated Chambers of Commerce and Industry of India report, points out that

> the mall culture has not been able to shift the focus entirely away from local traditional markets as the shoppers prefer to hang out and shop there, more so because of familiarity with ambiance, ease of access, variety of goods, loyalty of the customers.
>
> (*The Hans India* 2016)

The 'continuity' of periodic markets is neither unique to Delhi nor should it be seen as an aberration in a modern economy. Braudel (1983 [1979], 26) points that it is not possible to trace a linear history of the development of markets as 'the traditional, the archaic and the modern or ultra-modern exist side by side, even today'. Periodic markets exist in villages, towns, and cities across India and other parts of the world. When popular culture discourses and everyday conversations view weekly bazaars as central to the identity of Delhi, it should not be read as a factual statement of these bazaars being unique to Delhi or being more significant than other retail spaces. What such statements draw attention to is the ubiquitous presence of these bazaars in the lives of Delhi's residents. When a place is considered as central, unique, or special, it may not have anything inherently to do with that place but because one has a relationship with that place, a point I will further elaborate in the next section.

While the *form* of the weekly bazaar may be considered 'traditional' when compared to new spaces of consumption, it is not as if these bazaars have remained unchanged. One of the most obvious ways in which weekly bazaars continuously transform, and are thus *places in making*, is through their commodities. This is related both to rhythms of weather, festivals, and wedding seasons in the city and to larger processes that lead to commodities being replaced or added on, changing fashions and fads, and dynamics of the economics and technologies of materials. Commodities sold at the weekly bazaars of Delhi include mass-produced commodities sourced from different wholesale markets in Delhi (produced all over the country and beyond, especially China) or directly from factories in Delhi and other parts of northern India, as well as those produced in workshops and smaller factories in the informal sector. Counterfeit and stolen fast-moving consumer goods (FMCG) of multinational brands are also commonly available in weekly bazaars. Weekly bazaars are thus linked to various modern and not-so-modern, national and global chains of production and distribution. Part of their popularity is linked to the fact that they make available some 'old-style' commodities and services that are no longer easily available, e.g., commodities such as iron implements, grinding stones, and particular spices, and services such as knife sharpening. But a large part of their appeal also lies in their ability to continuously keep their commodities 'up to date', with the circulation of what is referred to in the bazaars as 'kya chal raha hai' ('what's trending') moving from customers to the bazaar and vice versa, and various other routes.

Why neighbourhood matters

Irrespective of the fact that commodities *within* weekly bazaars have not remained fixed, weekly bazaars *across* neighbourhoods do offer the same mix of wares and have overlapping vendors. These factors, along with the materiality of these bazaars, i.e., the use of aluminium and bamboo poles, tarpaulin, gunnysacks, cots, and rickety tables, may make it appear as if these bazaars are all the same. Why or how, then, does the neighbourhood matter?

Let us look at three ethnographic vignettes around the practices of three customers in the weekly bazaars they frequent in their respective neighbourhoods.

Sheenu Chawla is the owner of a popular beauty parlour in Bhogal, an urban village, later developed as a resettlement colony post-Partition, and she lives in the adjoining posh colony of Jangpura Extension. The Tuesday bazaar in Bhogal covers the street just opposite her parlour. She tells me that while she is not a regular shopper at the bazaar, if she has to go to a party or a social gathering on a Tuesday straight from the parlour, she often goes and buys a top from a stall in the bazaar, which is just a few steps away from the parlour. The vendor stocks the latest fashion in 'Western wear', and she can quickly pick up something for as cheap as Rs 150. 'Whenever I am wearing a top that I have bought from the bazaar, I get a lot of compliments. No one needs to know that I have bought it from the bazaar', she tells me. For Sheenu, buying imitation jewellery or 'smart tops' from stalls in the weekly bazaar is not about deception but about going against the snobbish belief that only expensive stuff is good—it is about being a smart consumer. She uses the term *samajhdari* (being wise) for someone who looks for and uses cheaper alternatives, and *bewakoofi* (being stupid) for those who spend extravagantly on frivolous things like clothes and jewellery. However, if Sheenu has to shop for a wedding, she always buys clothes from a big shop in Lajpat Nagar.

Shashi Kapoor, a retired army official, lives in Dwarka, a model 'planned' township in Delhi's suburbs, and goes to one of the several weekly bazaars in the neighbourhood every week. He is a member of a citizen's group called Dwarka Forum, which has been petitioning various local authorities to regulate the weekly bazaars in the area. The weekly bazaars are the main source for fresh fruits and vegetables in the locality. This is because, as he and other members of the forum jokingly tell me, the master plan of Dwarka 'forgot' to earmark an area for a permanent fruit and vegetable market. He tells me that it is the younger generation that frequents malls to buy vegetables: 'They come back late from office and then go to the mall to buy vegetables. But that does not work for our generation because the mall has rotten vegetables. Also, who buys vegetables from the mall?' When I ask him if he explores the bazaar, he says, 'I never roam around the entire bazaar. I take about 15–20 minutes, at the most half an hour. I know exactly the two–three vendors from whom I need to get vegetables and fruits from, and I buy my stock for the entire week.'

Yogesh lives in a low-income group Delhi Development Authority (DDA) colony in east Delhi, and works at a non-governmental organisation (NGO). He loathes shopping and leads a frugal lifestyle out of choice. But he likes roaming around the Monday bazaar in his neighbourhood. A visit to the bazaar usually includes a snack of 100 grams each of jalebi and pakora. He says the bazaar has *raunak* (literally, liveliness, beauty, or effervescence). Even though Yogesh loathes shopping, he likes picking up little gifts for his daughters from the bazaar—Barbie and Hannah Montana stickers, smiley fridge magnets, colouring books, Angry Birds pencil boxes. When his wife and kids, who live in Lucknow, come to visit, the Monday bazaar forms part of a family 'outing', along with India Gate and Janpath Market. On days that he does not feel like going down to the bazaar, he enjoys watching it from his terrace. 'At night, the lights of the bazaar look beautiful', he says.

These vignettes cannot, and are not meant to, capture the entire range of practices, affective ties, and routines that people build around weekly bazaars. They do, however, draw attention to how the same sort of place—the weekly bazaar—is enacted in different ways through the singularity of relationships that customers forge and renew with their respective bazaars. These experiences are mediated by class and gender, among other identities, which are themselves not fixed, as with specific needs (e.g., fresh vegetables) and routines (e.g., provisioning or leisure), creating different values that become attached to the place (smartness, freshness, beauty, etc.). These fabricated and shifting characters, identities, and values of places have been highlighted by Doreen Massey:

Individuals' identities are not aligned with *either* place *or* class; they are probably constructed out of both, as well as a whole complex of other things ... this
applies to places too. They do not have single, pregiven, identities in that sense.

(Massey 1994, 137, original emphasis)

If the weekly bazaar, like other places, is not pregiven or fixed, does this
mean then that the neighbourhood in which it is located is incidental? Is the
neighbourhood then just a site in which the bazaar, its objects, and relationships
are 'located'? The habits, practices, and routines of individual customers,
different and singular, through which the weekly bazaar becomes a place in
making or a place with different identities are possible due to the coming
together of two kinds of repetition—the weekly recurrence of the bazaar and
customers repeatedly frequenting the bazaar. Central to this intersection of two
repetitions is the proximity or co-presence of the bazaar and the neighbourhood.
While the neighbourhood is also fabricated through different associative milieus
(discussed in detail in the next section), it makes it possible for people to have
different levels and kinds of enduring relationships with the bazaar that is there
in 'their' neighbourhood. A recurrent theme in my conversations with
customers of different bazaars is how the bazaar they frequent is unique or
special, or even the best. Sometimes the distinctive features they pointed out
would vary, linked to their specific routines, preferences and tastes, or particular
objects, but often they would be identical in content, for instance, 'I get the
best bargains there', 'the vegetable vendors there don't charge for coriander and
green chillies', 'you get everything there', 'it's easy to navigate'. This brings us
back to the argument made in the previous section that what makes a place
unique for an individual may not be tied to an essential characteristic of the
place but rather to the person's *relationship* (built, reinforced, and modified
through familiarity, continuity, and repetition) with the place. Some of these
places you 'go to' or seek out as part of your personal itinerary, while others
you end up having a relationship with because they are close at hand (the
neighbourhood park, the street outside your house or office, the weekly bazaar
in your locality). The neighbourhood itself is fabricated through different
associative milieus, it is connected with other places, and its boundaries are not
fixed, as we will see in the next section, taking the case of Nizamuddin Basti.

'Everyone knows Nizamuddin'

This quote from what is considered one of the best guides and popular history
texts on Delhi, *Delhi: Its Monuments and History* (Spear 2006 [1943]) refers to
the locality in Delhi that derives its name from the much-revered Sufi saint
Hazrat Nizamuddin Auliya (1236–1325), and not the saint himself. Spear (2006
[1943], 37) writes, 'Everyone *knows* Nizamuddin [emphasis mine]. Very likely
you have been there already. Perhaps you went to a *mela* there. Or perhaps you
have seen it from a distance.' An annotation added to the text later (in 1994)
more specifically delineates the place being referred to as Nizamuddin Basti,

clarifying that the term 'basti' means an 'inhabited quarter' (p. 37, n1),[2] perhaps so that it is not confused with the colloquial, often pejorative, connotation of the term as a slum. This clarification, that the Basti is 'a settlement', 'an old village', or 'a continuously inhabited area', is oft repeated in walks conducted in Nizamuddin Basti by heritage experts and agencies, in books on Delhi, as well as by Basti residents, emphasising the historicity of the place, making possible different, sometimes conflicting, claims on the Basti.

The 'popularity' of Nizamuddin Basti has only grown over the years since the time Spear first wrote his book—as a preferred locality for Muslim migrants to settle in,[3] as a must-visit place on the tourist and heritage itineraries of Delhi, as a sacred geography centred around the saint's shrine, as a centre for the activities of the Tablighi Jamaat,[4] as a place to consume objects and experiences (e.g., qawwali, kebabs, biryani) fetishised as 'Muslim', as a historical site that needs to be 'preserved',[5] and in other ways. Due to these multiple *associative milieus* through which the place is constructed, one can argue, more so today, that *everyone* (not only residents of Delhi) *knows* the Basti. Each kind of knowing can, however, be unaware of; push to the background or hide; enable, build on, or add to; or be in confrontation with other kinds of knowing. These are not simply different representations of or discourses on the Basti but rather experiences and relationships enabled by various combinations of materials, artefacts, objects, affects, spaces, movements, and people. The relationships and experiences, in turn, make the Basti a *place in process*, continuously reconfiguring the boundaries and hierarchies between people and spaces. This is what Massey (1994, 154) refers to when she argues that every place is constituted from 'meeting places', that is, it is 'constructed out of a particular constellation of social relations, meeting and weaving together at a particular locus'. McCormack (2013, xi) makes a similar conceptualisation by suggesting that places need to be understood 'in terms of their enactive composition through practice' as 'affective spacetimes'. There is thus nothing essential or fixed about any place—it can be understood only through the various relationships to it, enacted in different associative milieus.

The multiple ways in which the Basti is made cannot fit neatly into de Certeau's (1988 [1984]) categories of 'strategies' and 'tactics', where the former is located in space and the latter in time. By strategy, de Certeau refers to the spatial ordering by powerful interests and by tactic, the appropriations and subversions by ordinary citizens, which undermine the former. The problem with the strategy–tactic dichotomy is that it does not account for the powerful and the disenfranchised using both strategies and tactics.[6] With much of Delhi's construction, including 'world-class' development projects, violating some land use code or building by-law (Ghertner 2015), the strategy–tactic dichotomy is not very useful in understanding place-making in the city. It is instead important to look at how different space-times and different kinds of place-making (everyday, imagined, affective, discursive) are hierarchically interconnected. The strategy–tactic, modern–vernacular, formal–informal dichotomies assume that one kind of space is given (or imposed) or defined, and is then appropriated. But often, as is particularly glaring in the context of Nizamuddin Basti, the 'official' or the 'dominant' spatial ordering is itself not fixed.

Nizamuddin is the name of a municipal ward with 19 residential colonies, including two that share the name Nizamuddin: Nizamuddin East and Nizamuddin West. Both are 'planned resettlement colonies', created for the influx of Punjabi migrants post-Partition, and are among the upscale neighbourhoods of Delhi. Nizamuddin Basti, a densely populated, low-income locality, is officially a part of Nizamuddin West, and is not demarcated as a separate locality in DDA's Zonal Development Plan (Gusain 1999). While addresses on the signage of shops and other structures in the Basti include Nizamuddin West, the terms 'Nizamuddin West' and 'Nizamuddin Basti' are rarely used interchangeably. The contrasts between the two neighbouring localities in terms of their built structures, public infrastructure, and atmospheres could not be starker. Nizamuddin West has well-maintained roads and parks and 'modern' houses, as opposed to the congested alleys, dilapidated buildings, and dusty parks in the Basti. So while the Basti might *be in* Nizamuddin West, it *is not* Nizamuddin West—the two are divided as much by boundaries of class as by their material and aesthetic appearances. This is despite Nizamuddin West becoming more predominantly 'Muslim' over the years, with Hindu residents selling off their houses to affluent Muslims (Jamil 2017, 87).

It is the 'look' of the Basti (narrow alleys, dense and narrow facades of houses, etc.) that makes middle- and upper-class residents of Delhi (including those of Nizamuddin West) label the Basti as a slum, unauthorised colony, or village. The administrative category of villages in Delhi is exempt from building by-laws and zoning regulations. Because the built structures of the Basti 'look' as if they do not conform to regulations, many NGO employees in the Basti have categorically and confidently told me that 'of course, it is an urban village, what else could it be?' The Basti, however, does not feature in either the Delhi government's or the DDA's list of villages. Nor does it appear in any official list of slums or unauthorised colonies. Some scholars (Datta et al. 1990) argue that this ambiguity around the Basti's 'official' status is deliberate and reflects a communal and class bias in Delhi's urban planning; for instance, if it were clearly a village in official records, its residents would be entitled to subsidised electricity for manufacturing and commercial activities. On the other hand, the Basti's construction as a 'historical village' gives the authorities and its partners, in the context of the urban renewal project of which the Basti is a part, greater control over the entire area rather than just discrete historical monuments.

But because the Basti 'looks' like a slum, it is deemed illegal not just by 'outsiders' but even by those considered part of the 'community'. In her book *Performing Heritage* (2012), Navina Jafa, who 'curates' walks in the city, quotes Khwaja Hasan Nizami, caretaker of the shrine of Hazrat Nizamuddin, from a walk she conducted for a group in Nizamuddin Basti:

Today, the *basti*, marked by large number [sic] of people, has become congested and is quite dirty. Large numbers of inhabitants here do not know or even care to know the rich cultural heritage that surrounds them. Many people have illegally occupied areas around the shrine and the neighborhood and have built illegally, sometimes at the cost of heritage buildings.

(Jafa 2012, 105, original emphasis)

The 'curator' gets Khwaja Hasan Nizami to speak to her group as part of her endeavour to make the walk more 'authentic' by bringing in a 'community representative'. Curated walks fabricate a certain experience of the Basti—an experience centred on monuments of historical and architectural significance, many of which have been restored or are in the process of being restored as part of the renewal programme. In such 'museumised' experiences (Jamil 2017, 54), the Basti becomes incidental or a hindrance to that experience, an experience that would perhaps be better without the lives inhabiting the place. What makes the quote significant is that it also brings to the fore the boundaries and frictions within the 'community'. Here it is important to note that the caretakers of the shrine, the *pirzade* family, have a stronghold in the Basti. They own large properties in the Basti and the adjoining Nizamuddin West; certain historical structures within and outside the dargah complex are under their control; and they claim to be the most original inhabitants of the Basti. Among the low-income residents of the Basti, too, there are divisions based on income, length of stay, and type of housing (e.g., kaccha versus pucca) among other factors, which are evoked in forms of claim making over the neighbourhood. Shared inhabitation of a place does not automatically or necessarily translate into a neighbourhood community.[7]

The official ambiguity around the Basti's status, on the one hand, makes parts of the locality prone to the possibility of being declared illegal and its inhabitants to eviction. But the same ambiguity, on the other hand, allows residents to claim themselves as its legal inhabitants and as being part of the historicity of the village. Not all encroachments on or within heritage structures are easily or successfully removed, with residents claiming these as homes (over generations) and arguing that their dwellings are legitimate because they have electricity and water connections (see, for example, Verma 2016). What could be a heritage structure in one associative milieu could be a home in another. That not all the people of the Basti value its cultural heritage, as Khwaja Hasan Nizami bemoans, does not mean that they do not have other kinds of connections with it, for instance, connections of making a livelihood by selling commodities or offering services to visitors to the Basti. Nor does it mean that they do not have a relationship with the Basti, for instance, through kinship and friendship networks built over the course of living in the neighbourhood. This illustrates that a place cannot be taken as given—it needs to be understood in terms of the experiences, uses, and values enacted in its making in different associative milieus.

What's so special about the Monday bazaar?

While it is important to understand how heritage discourses and practices are transforming Muslim neighbourhoods into '"museumized" space[s] for curiosity' (Jamil 2017, 85), it is also important to understand how these 'museumised', 'beautified', and 'renewed' spaces are re-incorporated into the everyday through the carrying on of lives within and around these spaces. And what are the other loci of different kinds of place-making that happen as lives are lived? The 'poor' are not just engaged in 'appropriating' spaces or resisting the 'appropriation' of their spaces. They also live lives through routines of work, domesticity, play and pleasure, shopping, meeting friends, lovers, and kin, etc., all of which are made possible by places and, in turn, make those places. One ordinary locus through which place-making happens in the Basti is the weekly bazaar held there every Monday. It is not one of the more public ways, discussed at the beginning of the previous section, in which the Basti as a place is known, studied, or debated, despite its centrality in the lives of women of the Basti. It is not only women's labour in homes and places of work that has been made invisible but also their role in the making of public places. The discourse around public places in Delhi being masculine and unsafe for women tends to exacerbate this invisibilisation. The following sections look at some of the ways in which the 'ordinary' Monday bazaar becomes significant, even special or famous, through the relationships that the Basti's women have with it.

Shape of the bazaar

The Basti is not a neutral site or terrain in which the Monday bazaar is located. The Basti and the bazaar are entangled in each other's associative milieus of place-making through their materialities and the routines and lives of the Basti's women. Shop designs as well as the layout of the bazaar are shaped by the character of the space in which the bazaar is located. Unlike most other weekly bazaars in Delhi, which are usually located alongside wide roads and/or pavements, the Monday bazaar meanders through the narrow labyrinths of the Basti. Most shops in the bazaar are on the ground level and make use of the walls lining the Basti to display wares. The lack of space does not allow for the more elaborate stalls you will find in some of the bigger weekly bazaars. Not only the shops but also the architecture of the bazaar moulds itself in the shape of the gullies of the Basti, and it is this that makes the bazaar accessible, usable, and enticing to the women of the Basti. As one of them explained why the bazaar is so popular among the women, 'It is for "ladies", lots and lots of women go [to the bazaar] because it runs alongside the park and is right in the midst of people's houses.'[8]

The women of the Basti do not passively consume the bazaar and its objects. They actively animate it through their repetitive practices in the bazaar. Rani, a resident of Nizamuddin Basti, considers herself an expert in what she refers to as 'bazaar karna' (literally, doing bazaar). It is by frequenting the Monday bazaar in the Basti for the last 18 years that she has mastered her techniques of bargaining and navigating the bazaar. She says,

> I have learnt how to shop. If someone [a vendor] sells a "suit" [salwar kameez] for 1,000 rupees, it's not like I will bargain and take it for 800 rupees. If someone says 1,000, then I make sure he brings it down to 500—I ensure that the price comes down by half. If I had to buy an expensive suit, wouldn't I just go to a shop [in a permanent market]?

Rani and others like her make and remake the Monday bazaar as a 'women's market' by frequenting it every week as a ritual. The bazaar has a large number of shops selling cloth, laces, buttons, thread, and sewing equipment to cater to the needs of the Basti women who prefer to sew their own clothes and many of whom sew for a living. The commodities in the bazaar not only make the Basti's women think that the bazaar is for them ('hamare liye') and theirs ('hamara bazaar') but also attract more women, now increasingly even from outside the Basti, reproducing the 'look' and reputation of the bazaar as a 'women's market'.

Rhythm of periodicity and proximity

The women of the Basti, like customers of weekly bazaars in other localities, have a diverse range of routines, familiarity, comfort, and affective ties with the Monday bazaar. But there is a heightened intensity about the excitement, attachment, and dedication of the Basti's women towards the Monday bazaar, which is enacted in their bazaar itineraries as well as their narratives of 'doing bazaar'. In conversations with them, they would repeatedly tell me that every woman goes to the bazaar, that they wait for Monday to arrive, and that they never miss it. Here, it is important to note that there are some women who rarely go to the bazaar, either due to financial difficulties or because their husbands or families do not allow them to. For the younger generation, boys and girls alike, 'doing bazaar' involves a very different experience because they are often running errands for their parents and have little freedom over the money and time spent in the bazaar. For women who are a part of and invested in the associative milieu of the bazaar, the bazaar forms such a significant part of their weekly routines in the Basti that they find it difficult to imagine someone not being part of it, hence their construction of the bazaar as a place where all women go. The periodicity and proximity of the bazaar make it possible for 'bazaar karna' to become a ritual for so many women. Saira tells me that

> Women of all ages go … men also go … they come with their wives … they carry the bags and their wives fill up the bags … It takes place every eighth day, anyone can wait for that long. They [women] think that the

Monday bazaar will anyway be held and I can get things from there. There is no other bazaar here close by … either we have to go to Bhogal, but then you have to shell out 40 rupees for the autorickshaw fare … so what's the gain … that's why women instead of going there wait for eight days to buy the things that they need.

Rani clarifies that she only buys what she needs instead of hoarding, 'but I go every Monday for sure', she says, even as her husband taunts her about it. 'Sometimes I just check out the shops, just in case I find something I like. *All women here go to the Monday bazaar*' (emphasis mine). On the one hand, she tells me that she does not go when she does not have money. On the other hand, she says, 'When I don't have money, I go do a quick round'. Sometimes she borrows money to 'find a way to go to the bazaar'. On the rare occasion that she does not go the bazaar, she looks at it from her terrace, which provides an excellent view of the bazaar. 'And then I hear of some good deal, and I am tempted to go down', she tells me.

The rhythm of the bazaar gets entangled with the rhythm of the Basti. On bazaar day, women usually wrap up their domestic chores quickly and those who work in the Basti itself either do not go to work or turn up for work late in the afternoon. A number of NGO supervisors working with women in the Basti told me how they would never schedule meetings for Mondays because no one would turn up. Noorjehan points out, 'I live in a joint family. Every Monday, the house is eerily quiet and empty. You see, in this market, you get good stuff, cheap stuff. So lots of people go. Everyone in my family goes.' The routine of going to the bazaar marks a break, perhaps even respite, from the routines of the everyday. Lefebvre (2004, 16) argues that there is 'no rhythm without repetition in time and space, without reprises, without returns, in short without measure. But there is no identical absolute repetition … there is always something new and unforeseen that introduces itself into the repetitive: difference.' Not only do different women do different things in the bazaar but also every visit is unexpected—you bump into friends and relatives you have not seen in a while (especially those from outside the Basti), you discover new vendors, new deals, new things, some of which are available only on that day, which is why it is imperative for women who are a part of the bazaar's associative milieu that they go every week.

Laughter and loyalty

The Basti's women often explain their ritual of 'bazaar karna' in very instrumental terms: the need to provision for home, to take advantage of good deals, to avoid wasting money on travelling to some other market, and to save money. But the manner in which they talk about the bazaar—eyes lighting up, talking excitedly, giggling—does not contain this instrumentality. This was heightened when I would speak to them in groups. There would be a lot of excitement, laughter, cracking of jokes, banter, conversations on the side, and sometimes even forgetting my presence. This atmosphere, which you also *feel* at the bazaar, between vendors and customers, is what makes the affective ties of the women to the bazaar, the full extent of which you can never really grasp unless you are part of that associative milieu of place-making. A very common pattern of banter that played out in front of me was when one woman would say, 'I don't really go every week; I only go when it is necessary', and someone else would interject saying, 'Arre, but I see you there every week', and then the group would laugh uncontrollably.

Sometimes, loyalties to the bazaar would be tested in these group conversations. In one of my interactions with a group of women, there was a playful episode of bickering between two of them. It began with Mehrunissa telling me that she does not go to the Monday bazaar. 'I don't like', she tells me curtly in English. Mehrunissa lives in Mayur Vihar in east Delhi, where she moved after her marriage. She visits the Basti because she has family there and also works there on and off. Before I can ask her why she does not like the Monday bazaar, Rani, who seems to have got the shock of her life, asks Mehrunissa, 'You really don't go?' 'No', says Mehrunissa. 'Really?' Rani says, still in disbelief. I ask Mehrunissa if she has never been there. 'I am a daughter of this place [Basti]', she tells me, so of course she has been to the bazaar; she grew up here and her school was right where the bazaar is held. I ask her why she does not like it then. 'It's nice. Everything about it is nice. Even the rates are good. But I prefer the Bhogal Tuesday bazaar', she says. At this point Rani can no longer contain herself and she asks Mehrunissa tauntingly,

> Here there is a 'facility' for everything. Why do you go to Bhogal? If I buy a kilo of potatoes, I can ask the vendor to give me green chillies for free without any hesitation, which is not the case in Bhogal.

After some time, Rani interrupts my conversation with Mehrunissa again because she has not got an answer yet.

RANI [R]: What's so special about Bhogal?
MEHRUNISSA [M]: Only Muslims come here [referring to the Basti] …
R: No, everyone comes here … they come from all kinds of places.
M: I can eat *gol gappas* [a snack] there. [Lots of laughter.]
R: But you get *gol gappas* here too.
M: There are lots of nice shops there. The market there is very nice and it's along the bazaar. You get everything in the market, from furniture to gold jewellery.
R: What's special about that? It's just shops.

The bickering continues till Rani is finally able to extract the real reason behind Mehrunissa's 'disloyalty' towards their bazaar and it turns out that a pickpocket had slit her bag. No money was stolen, but her bag was ruined and she stopped 'liking' the bazaar after that. Rani, who had a retort for everything, says 'So what? It's happened with me too. But it's got nothing to do with the Monday bazaar. It happens in Bhogal too. It happens in all crowded places!'

In the lives of the Basti women, the rhythms of the bazaar and the Basti are intertwined through their practices of 'doing bazaar' and so are their relationships to both places. Rani, the die-hard bazaar loyalist, defends the Monday bazaar not in terms of what is unique about it as opposed to other bazaars, say the one in Bhogal, but by arguing that when you can get everything right here, why go elsewhere? Mehrunissa, in turn, defends herself by clarifying that, even though she no longer lives in the Basti, her preference for Bhogal should not be misconstrued as her not having ties with the Basti.

'Now it is famous'

While the bazaar is central to the Basti women's routines of provisioning and pleasure, it should not be misconstrued as something that keeps them confined to the Basti. They go to other markets as well, for various reasons. As Noorjehan told me,

> I don't go to Bhogal often, but there are some things you get only there. For instance, there are many stationery shops there from where you can pick up things for the children. You also get 'gents' things from there; this [the Monday bazaar] is a 'ladies' bazaar. I go to Lajpat for shopping for Id, weddings and special occasions. Chitli Qabar is for buying expensive clothes for the children for a wedding or festival. You also get very nice *lehenga*s and saris there.

All these other places are also significant, but they do not involve the same degree of frequency and intensity that their interaction with the Monday bazaar involves. The Monday bazaar is also one of the few places where women go unaccompanied by anyone else, where their time spent is unaccountable.

While other markets or bazaars become important for 'special' shopping or shopping for something specific, many women drew my attention to how, over the years, the Monday bazaar has expanded its range of goods and its size, bringing itself closer to people's homes. Saira tells me,

> There has been a lot of 'changing' in the bazaar. Now it has become famous. Earlier it had just a handful of shops … you also got fewer things. Now you have shops for everything. That's its biggest benefit for people. The most significant 'changing' is that now the shops are set right in front of your door.

The other significant change is the expanding customer base of the bazaar to beyond the Basti women. As Saira explains,

> It's not as if only the poor shop here … lots of women who own boutiques come here … because the buttons and laces you get here are cheap. They

buy cheap from here and use it to sell expensive things. So see how useful the bazaar is for women!

Things that vendors stocked to cater to the needs of the Basti's women are now attracting others from outside the Basti. While Saira sees this as being beneficial to other women (of well-to-do classes), not all women in the Basti are so charitable about it. They point to the differences between their shopping practices and that of what they refer to as 'hi-fi' (high-class) women who do *bhaara-bhaari* (hoarding), which has pushed up prices of certain things and led to some things running out of stock quickly. 'Many of these women come early morning; they know exactly which vendors to go to and they will buy the entire stock of cloth', points out Noorjehan. (Even if the Basti women adjust their daily routine on bazaar day, most reach the bazaar by 10.30 am at the earliest because they first have to finish some chores at home.)

The Basti women, on the one hand, take pride in the increasing popularity of the bazaar, which they believe has led to greater variety in the things available to them, things that were earlier accessible only to a certain class. But that these other classes now come and buy things in bulk from their bazaar is something they look down upon. Noorjehan explains these changes in the bazaar and its shopping practices,

> Earlier you only got cheap things. Now you get the cheapest and the most expensive things. Now you get things from the 'export line', those things with a little bit of 'reject' [the reference is to export rejects, which are slightly defective]. So you get very nice clothes for children, ladies *kurtis*, shoes, sandals, etc., for very cheap. You will not get these things even in a shop in Bhogal; you will not even get it in Fab India [an Indian brand] ... Now the bazaar has become so 'popular' that women from Lajpat Nagar, those from 'hi-fi' colonies ... Punjabi women from Bhogal, who 'belong' to good families, well-to-do families, all these women come here now. What do the women of the Basti buy? They buy according to their need. Sometimes, if they really like something, then they borrow money for it. But these women who come from outside, they come with huge bags, and they keep stuffing things into those bags.

The distinction that the Basti women draw between themselves and 'other' customers is more to do with the nature of shopping practices strongly located in class distinctions. They do not make this distinction when it comes to others like them, that is, of similar class background, not religious identity, who come from outside the Basti to the Monday bazaar. These include their relatives, women who have moved out of the Basti, and women from neighbouring as well as far-off localities. Of these, some of the women from neighbouring localities (e.g., the slum near Oberoi Hotel) also visit the Monday bazaar regularly, while others come to the bazaar with varying frequency. The Basti women think that the reason 'their' bazaar gets so many visitors from near and far is because the things in the Monday bazaar are very good, much better and cheaper than those in other bazaars. But they also see it as a shopping practice that is similar to theirs, as Rani explains, 'Just as we are from Nizamuddin and we go to Seelampur for shopping, they also come from there.'

Conclusion

The Monday bazaar allows for everyday life in the Basti to continue through the very ordinary practice of provisioning, a practice that involves love, care, planning, and skills, which the Basti women continuously perfect through the repetitive practice of 'bazaar karna'. But the bazaar also marks an interruption in the space-time of the Basti, providing reprieve from routines of work and domestic chores, the everyday uses of a space, and the tight circle of everyday exchanges. It provides a space for casual intimacies, laughter, fun, and pleasure, which are equally important for carrying on everyday lives. The materiality, proximity, and objects of the bazaar, which are shaped by the materiality and lives of the Basti, make the women think of the bazaar as theirs. The bazaar might be an ordinary place, but the women make it special both by thinking it is theirs and making it theirs, by continuously renewing their relationship with it by going every week. As one of them said to me about the practice of 'doing bazaar', 'Ek baar chaloo ho gaya toh khatam hi nahin ho raha' ('Once it begins, it just doesn't end'). For most women, their lives in the Basti are inseparable from the Monday bazaar, intertwining their affective ties to both places, where one sometimes becomes a metonym for the other, as we saw in the tussle between Rani and Mehrunissa. That their bazaar is something that is attracting people from beyond the Basti only makes it more special, despite their disapproval of what women of 'hi-fi' classes do in the bazaar. As someone told me, 'The bazaar has become "world famous".' This statement is perhaps apt for the dargah side of the Basti, where monuments and graves are concentrated, and which has a market catering to tourists, pilgrims, and visitors from all over the world. 'World famous' in the context of the bazaar is a reference to the bazaar drawing people from near and far, to it being an open and enticing place, making the place more than just ordinary.

That something as itinerant and impermanent as a weekly bazaar can become a place that is *of* the Basti, and not just *in* the Basti, underlines the significance of place in the everyday lives of people. The bazaar would not be a place without the various kinds of relationships that different people have with it. There is no place without relationships, but relationships are also created in and through places, even as those places are continuously reconfigured, even as they are linked to other places.

Notes

1   By doing so, his work reveals how the so-called elite-takeover of cities is 'more of an aspiration than an empirical condition' (Anjaria 2016, 29), also highlighting how the

modernist and vernacular spatial configurations are not autonomous domains, with one acting upon (or resisting) the other, but are entangled in each other through relationships of negotiation and compromise.

2 Whether Nizamuddin Basti has been continuously inhabited over 700 years is much debated. Some argue that historical texts (for example, Hasan 1922) that mention a settlement in the area are probably referring to the dwellings of the *pirzade*, descendants of Hazrat Nizamuddin Auliya and custodians of the dargah. Heritage discourses as well as residents of the Basti counter this claim by drawing attention to the many mosques, tombs, and ruins of walls dating from between the thirteenth and sixteenth centuries. More than the historical accuracy of 'continued inhabitation', what is important is how this assertion is used in different kinds of place-making in the Basti.

3 Post-Partition, some members of the pirzade family migrated to Pakistan and their houses were let out to Muslim migrants from north India. At this time, Muslim migrants from elsewhere in the city also settled in and around the Basti. Over the years, particularly starting in the 1980s and 1990s (periods of communal violence directed at Muslims), Muslims from different parts of the country (Bihar, Uttar Pradesh, Assam) and from Bangladesh have chosen to settle in the Basti, with the dargah and the existence of previously arrived migrants possibly adding to a sense of security (see, for instance, Jeffery 1979; Snyder 2010; Weigl 2010). Jamil (2017) argues that the rise of segregated colonies in Delhi is not only related to communal politics but also maintained by a disciplinary state, neoliberal processes of globalisation, and stereotypical representations of Muslims.

4 The Tablighi Jamaat refers to the most widespread international revivalist movement in Islam. Its international headquarters, Islami Markaz, is located in Nizamuddin Basti. For an understanding of the role of the *markaz* (literally, centre) at Nizamuddin in the transnational institutional regime of the movement, and its relationship with Sufism and the Nizamuddin dargah, see Mayaram (2000 [1997], esp. Chap. 7) and Pieri (2015). Between the markaz and the dargah, the Basti receives the largest number of Muslim visitors compared to any other locality in Delhi (Jamil 2017, 85). The dargah attracts an average of 11,000 visitors a day (18,000 if it is a Thursday, when qawwalis are performed) and about 250,000 visitors during Urs (AKDN 2015, 10).

5 Nizamuddin Basti and its surroundings have a large number of medieval monuments. In addition, the Basti has a high concentration of graves—of kings, nobles, poets, and unknown people—a community of the dead 'linked to the living through a shared desire for barkat (grace) from the saint' (Datta et al. 1990, 2488). The Basti is part of a massive urban renewal project, based on a public-private partnership model, known as the Nizamuddin Urban Renewal Initiative. The project started in 2007 after the Aga Khan Trust for Culture (AKTC) and Aga Khan Foundation (AKF) signed a memorandum of understanding with the Archaeological Survey of India, Central Public Works Department, and South Delhi Municipal Corporation. The project is aimed at conservation-based urban development, socio-economic initiatives aimed at improving the quality of life for local communities, and better public infrastructure facilities.

6 For a nuanced understanding of how strategy and tactic (or political society and civil society) are intertwined in the context of urban spatial practices, see Anjaria (2016) and Roy (2007).

7 While the residents of the Basti are predominantly Muslim (Hindus and Christians constitute about 10 per cent of the population and live in specific streets), it is far from being a homogeneous 'community'. The demarcations among the residents are along the lines of the length of their stay in the Basti (long-established migrants versus relatively new ones), place of origin (different states of India and Bangladesh), language, caste, mohallas within the Basti (which in turn are linked to occupation,

caste, income, etc.), and housing type. That people shift to different kinds of housing within the Basti and sometimes even move out and that there are new migrants coming in also means that these boundaries are not fixed.

8   All quotations from the women of the Basti are translations from Hindi. Quote marks used within these quotations indicate that the words or phrases were used as such in the original interview or conversation. The quotations are from unstructured interviews and conversations with women in groups, conducted at Insha-e-Noor—a self-help group of the Basti women engaged in making handcrafted products—and with individual women at their homes. This section also draws on many walks and direct observations in the bazaar over the fieldwork period.

References

Aga Khan Development Network (AKDN). 2015. *Nizamuddin Urban Renewal Initiative: Annual Report 2015*. www.akdn.org/sites/akdn/files/media/publications/2015_annual_re port_-_nizamuddin_urban_renewal_initiative.pdf. Accessed April 5, 2018.

Anjaria, Jonathan Shapiro. 2016. *The Slow Boil: Street Food, Rights and Public Space in Mumbai*. Stanford, CA: Stanford University Press.

Banerjee, Sumanta. 2016. *Memoirs of Roads: Calcutta from Colonial Urbanization to Global Modernization*. New Delhi: Oxford University Press.

Braudel, Fernand. 1983 [1979]. *Civilization and Capitalism: 15th–18th. Volume II: The Wheels of Commerce*. London: Book Club Associates.

Datta, Pradip, Biswamoy Pati, Sumit Sarkar, Tanika Sarkar and Sambuddha Sen. 1990. 'Understanding Communal Violence: Nizamuddin Riots', *Economic and Political Weekly*, 25(45): 2487–2495.

de Certeau, Michel. 1988 [1984]. *The Practice of Everyday Life*. Trans. by Steven Rendall. Berkeley, LA and London: University of California Press.

Donner, Henrike and Geert De Neve. 2006. 'Space, Place and Globalisation: Revisiting the Urban Neighbourhood in India', in Geert De Neve and Henrike Donner (eds), *The Meaning of the Local: Politics of Place in Urban India*, 1–20. London and New York: Routledge.

Economic Times. 2013. 'How Delhi's Street Markets Clock Rs 20–30 Lakh Turnover', June 3, 2013. https://economictimes.indiatimes.com/slideshows/biz-entrepreneurship/how-delhis-street-markets-clock-rs-20-30-lakh-turnover/slideshow/20408215.cms. Accessed April 7, 2018.

Escobar, Arturo. 2001. 'Culture Sits in Place: Reflections on Globalism and Subaltern Strategies of Localization', *Political Geography*, 20: 139–174.

Ghertner, D. Asher. 2015. *Rule by Aesthetics: World-Class City Making in Delhi*. New York: Oxford University Press.

Gupta, Akhil and James Ferguson. 1992. 'Beyond "Culture": Space, Identity and the Politics of Difference', *Cultural Anthropology*, 7(1): 6–23.

Gupta, Akhil and James Ferguson (eds). 2001 [1997]. *Culture, Power, Place: Explorations in Critical Anthropology*. Durham and London: Duke University Press.

Gusain, R. S. 1999. *Zonal Development Plan, Zone (Division—'D' New Delhi)*. New Delhi: Delhi Development Authority.

Hasan, Maulvi Zafar. 1922. *Memoirs of the Archaeological Survey of India, No. 10, A Guide to Nizamuddin*. Calcutta: Superintendent Government Printing.

Hashmi, Sohail. 2007. 'The Hafta Bazaars of Delhi', *Kafila*, August 6, 2007. https://kafila.online/2007/08/06/the-hafta-bazaars-of-delhi/. Accessed August 10, 2015.

Jafa, Navina. 2012. *Performing Heritage: Art of Exhibit Walks*. New Delhi: Sage.

Jamil, Ghazala. 2017. *Accumulation by Segregation: Muslim Localities in Delhi*. New Delhi: Oxford University Press.

Jeffery, Patricia. 1979. *Frogs in a Well: Indian Women in Purdah*. London: Zed Press.

Lefebvre, Henri. 2004. *Rhythmanalysis: Space, Time and Everyday Life*. London and New York: Bloomsbury Academic.

Massey, Doreen. 1994. *Space, Place, and Gender*. Minneapolis, MN: University of Minnesota Press.

Mayaram, Shail. 2000 [1997]. *Resisting Regimes: Myth, Memory and the Shaping of a Muslim Identity*. New Delhi: Oxford University Press.

McCormack, Dereck P. 2013. *Refrains for Moving Bodies: Experience and Experiment in Affective Spaces*. Durham and London: Duke University Press.

Pieri, Zacharias P. 2015. *Tablighi Jamaat and the Quest for the London Mega Mosque: Continuity and Change*. New York: Palgrave Macmillan.

Robinson, Jennifer. 2006. *Ordinary Cities: Between Modernity and Development*. London: Routledge.

Roy, Ananya. 2007. *City Requiem, Calcutta: Gender and the Politics of Poverty*. New Delhi: Pearson Publishing.

Shaw, Annapurna. 2012. *Indian Cities*. Oxford India Short Introductions. New Delhi: Oxford University Press.

Snyder, Michael. 2010. 'Where Delhi Is Still Quite Far: Hazrat Nizamuddin Auliya and the Making of Nizamuddin Basti', *Columbia Undergraduate Journal of South Asian Studies*, I(II): 1–29.

Spear, Percival. 2006 [1943]. *Delhi: Its Monuments and History*. Updated and Annotated by Narayani Gupta and Laura Sykes (in 1994). New Delhi: Oxford University Press.

The Hans India. 2016. 'Delhi's Swanky Shopping Malls No Match for Weekly Bazaars', *The Hans India*, June 14, 2016. www.thehansindia.com/posts/index/Life-Style/2016-06-14/Delhis-swanky-shopping-malls-no-match-for-weekly-bazaars/235121. Accessed April 7, 2018.

Verma, Richi. 2016. 'House for the Dead, Home to the Living', *The Times of India*, January 17, 2016.

Warf, Barney and Santa Arias. 2008. 'Introduction: The Reinsertion of Space into the Social Sciences and Humanities', in Barney Warf and Santa Arias (eds), *The Spatial Turn: Interdisciplinary Perspectives*, 1–10. London: Routledge.

Weigl, Constanze. 2010. *Reproductive Health Behaviour and Decision-Making of Muslim Women: An Ethnographic Study in a Low-Income Community in India*. Berlin: Lit Verlag.

Part II

Stories

Neighbourhoods as imagined
and narrated entities

4 Two tales of a neighbourhood

Eyüp as a stage for the Ottoman
conquest and Turkish War
of Independence

Annegret Roelcke

Introduction

Istanbul's Eyüp neighbourhood has, since the 1990s, been the object of numerous activities by various groups claiming to protect and promote the identity of the neighbourhood (Hammond 2016, 98–144; Öztürk 2017). Many projects relate Eyüp's identity to the shrine of the Prophet's Companion Ebu Eyyub, after whom the neighbourhood is named. The shrine, with its adjacent monumental mosque and square, constitutes a central feature of Eyüp's built environment. This common element notwithstanding, the narratives of various actors differ in their periodisation and framing as well as in their modes of authentication and communication. Referring to events in the past to characterise the identity of the place in both past and present, they narrate Eyüp as a stage for the Ottoman conquest of Istanbul in 1453 or for the Turkish War of Independence (1919–1923).

Comparing two institutions' narratives about Eyüp as well as their different ways of engaging with these narratives in the context of identity construction and power struggles, this chapter points to the 'palimpsest' nature of places embodying traces of 'many different times and histories' (Huyssen 2003, 94; Winter 2009), which makes possible a 'multiplicity of readings' of both their past and present and points to the volatile and processual character of their socially constructed identities (Massey 1995, 184–186). The chapter further demonstrates the multiple ways in which people engage with memories of the past, related to diverse motivations, possibilities, and restraints in contemporary socio-political constellations (Hodgkin and Radstone 2003; Nikro 2019, 18).

Images from two books indicate these different narratives and engagements with memories of the past. Both books juxtapose earlier and contemporary photographs of the same spots in Eyüp, but their choice of photographs and how they are presented construct different narratives about the neighbourhood's history. One is a publication by the Justice and Development Party (*Adalet ve Kalkınma Partisi*, AKP)-led Municipality of Eyüp, the other is a book edited by Şener Türkmenoğlu. The latter has been an active member of a group calling themselves 'Eyüp's Friends'. Since their institutionalisation as the Foundation of Eyüp's Friends (*Eyüp Dostları Vakfı*) in 2010, he has been a part of its administrative board, serving as vice chairman since 2017 (Eyüp Dostları Vakfı n.d.-a). His book was published by the Hazinedaroğlu Construction Company on the initiative of its vice president İlhan S. Ergelen, a major industrialist. Both Ergelen and Türkmenoğlu, as well as the main author Feridun Benden, grew up in Eyüp. The Municipality of Eyüp and the foundation are the main protagonists of this chapter.

Eyüp's Golden Years '94-'09 was published by the Municipality of Eyüp in 2008. It contrasts photographs taken before and after interventions into the neighbourhood's physical fabric by the municipality as led by the Welfare Party (*Refah Partisi*, RP), which took office there in 1994 (see Figure 4.1). The party was the predecessor of the AKP, which has been in power since 2002. The book describes two photographs of Eyüp's cemeteries as follows:

Tarihi Eyüp Sultan Mezarlıkları harap haldeydi

Restorasyon sonrası mezarlıklar

Figure 4.1 Eyüp cemeteries before (above) and after (below) transformations by the Welfare Party-led Municipality of Eyüp.

Source: Municipality of Eyüp. 2008. *Eyüp'ün Altın Yılları '94-'09*. Istanbul: Eyüp Belediyesi. Page 202. With kind permission of the Eyüpsultan Municipality.

We took the historical Eyüpsultan cemetery, that [...] had actually turned into a garbage dump, a place of the drunk and outcast, [...] under protection. We have surrounded the environs of the whole cemetery with high support walls, fitting to the original fabric. [...] We have in fact turned the historical Eyüpsultan cemeteries, by erasing its view, which has until recently been frightening to those who entered, into a flower garden.

(Eyüp Belediyesi 2008a, 202)

The book suggests the RP's taking office in 1994 to be a major turning point in the history of Eyüp (Roelcke 2020). The time before 1994 is associated with physical and moral decay. Since 1994, the municipality claims, Eyüp has been provided with modern living standards and its architectural heritage has been restored (Eyüp Belediyesi 2008a, 14). Claiming to have brought back 'original' features, the municipality proudly announces that it has 'erased' the cemeteries' recent past.

A Moment Comes, When Time Stops ... Eyüp (2007), edited by Şener Türkmenoğlu, describes the cemetery's architecture in a picture of Eyüp's Beybaba cemetery street taken in 1919 (see Figure 4.2): 'The iron railed windows [...] seem to soften the feeling of death! [... Humans] are not afraid of the hereafter' (Türkmenoğlu 2007, 34). The contemporary image of the same spot, on the other hand, is criticised bitterly: 'If there is a sin somewhere, for sure it is in Bey Baba Street's renewal. [...] the lines of replica work squeezing the spirit [...] It makes people feel lonely, sad, and frightened' (Türkmenoğlu 2007, 35).

The same kind of renewal work on the cemetery walls conducted by the current Municipality of Eyüp, which it praises as fitting to the 'original' and for increasing security, is harshly criticised in Türkmenoğlu's book as a 'sin', 'squeezing the spirit', and evoking fear of death. The present condition is evaluated negatively and contrasted with an earlier one portrayed as a spiritual harmony between humans, life, and death. This time is exemplified by a photo taken in 1919, during the Ottoman Empire's last years, when Istanbul was occupied by the Allied forces and the Turkish War of Independence started.

The two books differ not only in their evaluation of an urban renewal project, they also construct two different versions of the neighbourhood's history, introducing different actors and events as bringing about historical changes, which are in turn portrayed as marking periods of time characterised in very different terms. Physical structures and landscapes often centrally feature as representing historical narratives, seemingly proving them true as material traces of the past (Mills 2011, 187). However, the same objects and landscapes—in this case, stone walls in the Eyüp cemeteries—can be interpreted in various ways to support different and even contradicting historical narratives (Winter 2009; Hodgkin and Radstone 2003, 11–13). Various narrators in the present refer from their multiple perspectives in diverse narratives to places' pasts to construct both their identities and narratives about the present and the future. Historical narratives, understood as connecting an imagined past to present and future conditions, are often contested between various actors, related to competing

Figure 4.2 Beybaba cemetery street in 1919 (above) and in the mid-2000s (below).

Source: Recent image by Şener Türkmenoğlu. In: Türkmenoğlu, Şener, ed. 2007. *Bir An Gelir, Durur Zaman ... Eyüp*. Istanbul: Hazinedaroğlu İnşaat Grubu. Pages 34 and 35. With kind permission of Şener Türkmenoğlu.

interests in contemporary socio-economic power constellations (Massey 1995; Hodgkin and Radstone 2003; Trouillot 2015; Christophe, Kohl, and Liebau 2017).

Indeed, the two books represent different social groups with competing claims within local and national power constellations. The AKP-led Municipality of Eyüp's book celebrates its own accomplishments and the years since its predecessor took office as 'Eyüp's golden years'. It is therefore directly connected to legitimising its own rule. This narrative resonates with the general discourse of the AKP, which is usually understood as constituting an Islamic-religious narrative of society.

A Moment Comes, When Time Stops ... Eyüp was created by people connected to the group that formed the Foundation of Eyüp's Friends. The foundation represents a formerly dominant social group that has felt pushed aside by shifting class formations, especially since the rule of the AKP and its predecessors. They criticise current developments by juxtaposing contemporary photos with those symbolising earlier, supposedly harmonious times. Presenting themselves as knowing Eyüp's original condition, they claim prominence in the contemporary neighbourhood's socio-political landscape. Their narrative resonates with general narratives commonly understood as secular and Kemalist, and as opposed to the AKP's narrative.

In the following pages, I will examine the different historical and identity narratives of the AKP-ruled Municipality of Eyüp and the members of the Foundation of Eyüp's Friends, as well as their different ways of engaging with these narratives. Both are important players in local heritage production.

In order to discern the municipality's narrative, I rely on *Eyüp's Golden Years*, the target audience of which was Eyüp residents. The book provides a long list of the municipality's accomplishments since 1994. The second main source that I use are the municipal city guidebooks about the neighbourhood, which mainly target tourists. Both genres seek to promote a specific version of the neighbourhood's identity and the current government's relationship to it.[1]

Compared to the municipality, the foundation has only few official media disseminating its narrative. As such, I examine the foundation's website, which involves texts and a video documentary created about the foundation by the television programme *Kayda Değer*. I will also analyse the permanent exhibition of the foundation's museum, called Eyüp Historical Life and Culture Center, as well as the book *Eyüp's Story of the Last Century—From the Speech of the Living Ones* (2018) on Eyüp's oral history, the publication of which was officially supported by the foundation. Şener Türkmenoğlu is a central character connected to these sources. He initiated both the museum and the oral history project and edited the book, and features extensively in the video. Also considering his prominent position in the foundation's administrative structure, Türkmenoğlu's activities and narratives seem to benefit from substantial support by the foundation's members. I take this as a justification to additionally base my analysis on three earlier books written or edited by him in order to have a body of sources more comparable in kind and size to the municipal books. These books were published during the formative period of the foundation, but before its official founding, and therefore without its official support. At that time, Türkmenoğlu was already actively involved in the group (Eyüp Dostları Vakfı n.d.-a). Along with the aforementioned *A Moment Comes, When Time Stops ... Eyüp*, these books are *Eyüp. Losing One's Heart to a Neighbourhood* (2005) and *History of Eyüp Sport Club in Its 90th Year and of the District's Clubs* (2009). They seek to document local history based on photographs, documents, and interviews. I also conducted an interview with Türkmenoğlu in order to find out more about the foundation's institutional structure and its activities, about which less information is publicly available than in the case of the municipality.

While a comparative presentation poses the danger of creating artificial dichotomies, a scholarly comparison in this context is interesting precisely because the actors themselves construct a difference between each other and the groups they represent. This takes place in numerous ways and sometimes only implicitly by the different actors, connected to their positions and varying opportunities to disseminate their messages in the current power structures (Kresse 2017). Thus, the municipality, being in a position of power, rather than addressing difference seems to try to involve the foundation's members in its own activities to give the impression of unity and support. The foundation's members do at times collaborate with the municipality. Yet, they simultaneously assert their difference to the currently dominant social groups and their narratives, and express opposition to current developments in Eyüp.

The two institutions' organisational structures and capabilities for public outreach differ greatly. One is a municipality with huge financial and organisational resources and the support of the nation-wide ruling party. The foundation, on the other hand, constitutes a local non-governmental organisation that functions to a large part via the voluntary work of its members, as well as membership fees and donations. However, they are comparable in the sense that they locally represent the two main groups commonly perceived to stand in polarised opposition to each other in contemporary Turkish society. These are the declaredly Islamic-religious and secular fractions. The two actors' institutional differences, along with their respective narratives and communication strategies, are intimately linked to the particular local constellation of these two groups within Eyüp. Comparing the two groups' narratives shows that the Eyüp neighbourhood is not homogeneous, nor does it hold to one message, as especially the AKP-led municipality's narratives suggest. Instead, Eyüp is, like other places, open to multiple different identity constructions. Second, examining the narratives of actors who position themselves as different from each other can also show similarities in their structures, and thereby point to more generally shared sensibilities. It is, however, important to keep in mind that the boundaries perceived between the two groups are not absolute, nor are their constructed identities homogeneous or stable. People may be linked in various formal and informal ways to one or both institutions and construct their own and the place's identity in multiple ways. Connections to the municipality can be especially diverse, due to its size and its multiple functions, i.e. being connected to the local

AKP branch, as a government, service provider, or employer. Instead of examining individual identity constructions, however, this chapter's comparison focuses on the narratives that are publicly communicated by the institutions and their representatives.

Sites of memory in Istanbul's cityscape

The Municipality of Eyüp's and Foundation of Eyüp's Friends' promotion of narratives on Eyüp's past can be viewed within a global trend of branding cities with 'cultural heritage' to attract investment and tourism. Since the 1980s, Istanbul and areas close to Eyüp have been reshaped by this objective, in which the promotion of Ottoman heritage has played a major role (Öncü 2007, 233–237; Bezmez 2008). The RP's and AKP's ascension to political office in Istanbul in 1994, and nationally in 2002, contributed to increased public engagement with the Ottoman past due to the parties' Islamic-Ottoman revivalist rhetoric. While they organised events glorifying Istanbul's and the country's Ottoman past, such as re-enactments of Istanbul's Ottoman conquest (Çınar 2001), people perceiving themselves as secular reacted with fear that the values of the secular republic would be lost. In what Esra Özyürek calls a 'nostalgia for the modern' (2006), they mobilised memory of the early republican era. Additionally, in the course of Turkey's negotiations for accession to the European Union, human rights for so-called minorities were reformed. Related to this, histories of the presence of and violence against Armenians, Greeks, and Jews, often denied in Turkish nationalist narratives, began to be debated in the 1990s and 2000s (Mills 2005, 446–450). Within this general context of increased public remembrance of the past, diverse narratives on the Ottoman past became popular in the 2000s, referring to it variously as a blueprint for multiculturalism and tolerance or for religious-nationalist projects. They differ from the mainstream Kemalist narrative that dissociates the Turkish nation from the Ottoman past (Fisher Onar 2015, 149–152; Öncü 2007).

Providing material traces for these narratives, the cityscape of Istanbul as the former Ottoman capital features prominently in many of them. Spaces for Ottoman nostalgia have been created, both through the provision of goods marketed as traditionally Ottoman, such as food or entertainment (Karaosmanoglu 2010), and through the restructuring of the physical fabric, which often involves older structures being demolished and residents displaced. Many times, these processes are related to economic interests and catering to more affluent consumers and tourists (Mills 2005).

Studies of the interpretation of Istanbul's cityscape and its physical restructuring along narrative lines demonstrate how the same neighbourhoods can be employed to support opposing identity narratives. Thus, Beyoğlu has been interpreted using its earlier non-Muslim population as symbolising Istanbul's European identity, or as an example of an Ottoman governance model of tolerance (Bartu 1999; see also Chauvel 2011). Studies also point to the complexity of nostalgic place-making by showing how elites draw on the past of former non-Muslim neighbourhoods for constructing their own identity as cosmopolitan and tolerant, but how in doing so they obscure past violence against non-Muslims and prompt new displacements via gentrification (Mills 2005, 2011). Thus, many studies on nostalgic place-making in Istanbul examine contemporary engagement with non-Muslim pasts by currently dominant groups with Muslim family backgrounds, perceiving themselves as secular or religious.

The narratives examined in this chapter, on the other hand, do not refer to non-Muslim groups to construct Eyüp's identity. Instead, they narrate Eyüp as a pure and ideal version of an imagined larger community (Anderson 1991), identified with a relatively homogeneous Islamic-Ottoman or ethnic Turkish character. They claim Eyüp played a significant and heroic role in ushering in major political developments in these imagined communities' pasts, and thereby claim its potential importance for future developments. Tolerance and cosmopolitanism rarely feature in their narratives. Instead, portraying themselves as the heirs to Eyüp's glorious heritage, the Municipality of Eyüp and the Foundation of Eyüp's Friends represent different social groups competing for dominance within Eyüp. In their narratives, both relate to the neighbourhood's particular socio-political constellation, as well as to competing identity narratives on a more general level (Bora 2003; Öncü 2007; Fisher Onar 2015).

Both the municipality and the foundation frame their activities in accordance with Eyüp's imagined historical identity, portrayed as being lost and in need of preservation or revival. They differ in their characterisation of this imagined glorious past and in their ideas on how to revive it. Rather than directly contradicting each other, their historical narratives create different historical moments or facts that they use to establish continuity, while silencing (Trouillot 2015) other processes (see also Nikro 2019, 8; Hodgkin and Radstone 2003). Their selection of what to highlight and silence, as well as their distinct framings and interpretations, lead to differing general historical narratives and evaluations of the neighbourhood's identity and current condition (Trouillot 2015, 108–140). Both the municipality and the foundation's members are in an ambiguous position as they are simultaneously involved as actors in the socio-historical process and in the construction of the narrative about this process. This points to the multiple ways in which power is connected to the making of history, both in its socio-historical and narrative forms (Trouillot 2015, 24, 29).

In order to authenticate and communicate their narratives, both the foundation and the municipality attempt to create what Pierre Nora has influentially coined *lieux de mémoire*, sites of memory, being 'embodiments of a memorial consciousness'. Nora contrasts these *lieux* as constructed 'by artifice and by will' to generate previously lacking social cohesion (1989, 12) with '*milieux de mémoire*, real environments of memory' (p. 7), which he describes as 'borne by living societies' (p. 8). However, as Norman Saadi Nikro criticises, this dichotomy 'underestimates how people proactively work on and experience memory as social engagement', and 'discards the value of memorials and commemorations as modalities of social exchange' (2019, 18). Indeed, the different media and means that the foundation and the municipality employ to authenticate and communicate their narratives result in numerous ways in which people actively engage with memory, both in the construction of sites of memory and in their appropriation, and by doing so, engage with each other. These ways, however, vary considerably in terms of the formats and spaces they use.

Socio-historical context

Several members of the foundation portray themselves as belonging to families that have lived in the neighbourhood for centuries. Located at the Golden Horn, outside the city walls, the neighbourhood of Eyüp constituted the centre of one of Istanbul's four administrative districts throughout most of the Ottoman period. It had a vast hinterland that provided the intramural city with agricultural goods. Containing numerous shrines, mosques, religious convents, and large cemeteries, the neighbourhood also functioned as a pilgrimage site. According to Tülay Artan, the economic situation of the majority of central Eyüp's inhabitants in the eighteenth century was rather modest, while waterside mansions were built on Eyüp's Golden Horn shore for the capital's elite, especially female court members (Artan 1998; Faroqhi 1998). Related to the military reforms by Mahmud II in 1826, Rami barracks were built on a hill close to Eyüp, around which settlements developed (Yenen, Akın, and Yakar 2000, 88). The Feshane factory was opened at Eyüp's Golden Horn shore in 1833 to produce headgear for the soldiers (Yaramış 2004). More factories were built at the Golden Horn during the industrialisation of the nineteenth century, which attracted workers to settle. The neighbourhood further grew with the arrival of refugees from the Balkan regions lost by the Ottoman Empire in the Russo-Ottoman War in 1877–1878 and the Balkan Wars in 1912 and 1913 (Yenen, Akın, and Yakar 2000, 88). Several members of the foundation also claim to descend from these families, thus having arrived in Eyüp during the Ottoman Empire's last years.

Industrialisation of the Golden Horn area and the concurrent, often informal settlement of workers continued after the foundation of the Republic of Turkey in 1923 and increasingly after the 1950s. In the 1940s, more migrants from the Balkans were moving to Rami, while the vast majority of those arriving after the 1950s came from Turkey's Black Sea region. The Golden Horn's deindustrialisation in the 1980s in the context of a general liberalisation of Turkey's economy massively affected Eyüp's economic base. While some residents left for new industrial areas, others started to work in the service sector. In parallel with efforts to clean up the Golden Horn, plans were made to protect Eyüp's historical fabric and to attract tourists (Yenen, Akın, and Yakar 2000, 88, 103, 119; Hammond 2016, 104).

The first three Eyüp mayors from the AKP or its predecessor party, Ahmet Genç (1994–2009), İsmail Kavuncu (2009–2014), and Remzi Aydın

(2014–2019), all belong to families who moved to Istanbul from Anatolia after 1950. Current AKP mayor Deniz Köken (since 2019) was also born in Anatolia and moved to Istanbul at the beginning of his career. Genç and Aydın spent their childhoods in Eyüp, while the fact that Kavuncu was not 'from Eyüp' caused discontent (Çakır 2016). The AKP and its predecessor parties have claimed to represent those who migrated from rural areas to cities like Istanbul. These people have often felt neglected by urban elites by them denying their belonging to the cities, and by lacking provision of infrastructure (Tuğal 2009). Nevertheless, Genç and Kavuncu were rather powerful businessmen, and Aydın is a lawyer who also studied Islamic theology. Köken is an economist who worked in the banking sector (Eyüpsultan Municipality n.d.[2]).

After the group calling themselves 'Eyüp's Friends' had met informally since 1989, the Foundation of Eyüp's Friends was founded in 2010. It aims 'to strengthen social life in order to provide unity, togetherness, mutual help and solidarity among the people of Eyüp (*Eyüplü*)' (Eyüp Dostları Vakfı n.d.-a). The foundation is located in an upper storey of a building in a backstreet of Eyüpsultan Square. Its major activities include providing scholarships and other forms of welfare, and organising seminars, 'cultural events', and meetings among *Eyüplüs*. Several of the foundation's founders occupied relatively high positions in the Turkish military and state administration or otherwise belong to the upper middle class. Some of the older members moved to more luxurious neighbourhoods when the number of informal settlements in Eyüp grew (Şener Türkmenoğlu, interview with author, 1 December 2016). The first meetings of the group were thus held in military officers' clubs outside of Eyüp; in Sarıyer between 1989 and 1995, in Kasımpaşa in 1998, and in Harbiye in 2008 (Eyüp Dostları Vakfı n.d.-a). The vast majority of the founding members were male and many were retired. In recent years, women have become more involved, especially in organising activities, while the management board's average age has dropped slightly. Many members seem to sympathise with social-democratic or Turkish-nationalist parties perceived as secular and not with the ruling AKP. Locally prominent figures of the Republican People's Party (*Cumhuriyet Halk Partisi*, CHP), the Nationalist Movement Party (*Milliyetçi Hareket Partisi*, MHP), and İYİ Party are members of the foundation (Eyüp Dostları Vakfı 2017[3]).

Eyüp as the stage of Istanbul's Ottoman conquest

The cover of a city guidebook about Eyüp published by the municipality in 2008 illustrates the main elements of the municipality's narrative (see Figure 4.3). The guide is a reprint of one from 1996 that had been directed at Eyüp's residents as a way for the RP to present itself to its electoral base two years after assuming power. The 2008 guide has a new visual design, including numerous photographs. It is bilingual, in Turkish and English, targeting mainly tourists. Further editions came out in 2011, 2012, and 2015. In 2016, the municipality published a new guide.[4] The history section in *Eyüp's Golden Years* (Eyüp Belediyesi 2008a, 16–17) is a shortened and only slightly amended

Figure 4.3 Cover of the Municipality of Eyüp's 2008 city guidebook.

Source: Municipality of Eyüp. 2008. *Eyüp Şehir Rehberi: The City Guide*. Istanbul: Eyüp Belediyesi. With kind permission of the Eyüpsultan Municipality.

version of the guidebook's text, which is generally the same in the editions from 1996 to 2015.

The 2008 guide's cover presents a collage of photographs of places in Eyüp, presumably taken recently. They are lit in warm colours and convey a festive and mystical atmosphere. The architectural structures depicted—Eyüpsultan Mosque, Eyüpsultan Shrine, and an Ottoman tombstone—point to the religious and spiritual significance attributed to the place. The shrine of Ebu Eyyub is claimed to make Eyüp 'the fourth most sacred Islamic pilgrimage place after Mecca, Medina and Jerusalem' (Eyüp Belediyesi 2008b, 67) and to have legitimised Ottoman political rule (Eyüpsultan Municipality Mayor's Office 1996, 1). Also, nowadays, politicians from different parties, the AKP prominent among them, perform prayers at the shrine before and after important political events such as elections (Acarer 2017).

Eyüp's architectural structures also represent cultural heritage ('*kültürel varlıklar*') and are listed within the guide (Eyüp Belediyesi 2008b, 150–263). Eyüp's cultural heritage is portrayed as Ottoman, as only Ottoman-era structures are shown. Ottoman is interpreted as Islamic, as only structures related to Islam are depicted on the cover.

A third element in the municipality's narrative is what is called 'nature', mainly relating to plants in the cemeteries and the water of the Golden Horn:

> The relationship between "Eyüpsultan" and "Islam", that began on the hillsides next to the Golden Horn and rested on the dark and quiet green sites, grew, blossomed and developed with its climate and with its plants fitting to this soft climate like a plant, which has found its soil.
>
> (Eyüp Belediyesi 2008a, 16)

The Municipality of Eyüp repeatedly stresses its taming and controlling of nature, which is otherwise portrayed as wild and possibly dangerous. Thus, in the example of the cemeteries, the municipality presents the graves not only as having been damaged by the encroachment of unkempt weeds before its interventions, but also as providing shelter to people pursuing immoral activities. In contrast, after the restoration, nature in the form of roses is portrayed as something beautiful and peaceful, as long as it is in a secured enclave within stone walls. Nature's beauty is portrayed in this way as a timeless and intrinsic quality of the place parallel to its sacredness and mystical atmosphere. It is a major element of the municipality's construction of Eyüp as a place of inner peace (*huzur*) in opposition to stressful modern life in a megacity (Roelcke 2019).

Eyüp's history is generally periodised in the municipality's publications as follows: The narrative starts with Prophet Muhammad predicting the Islamic conquest of Istanbul. The Prophet's Companion Ebu Eyyub then dies around the year AD 669 in an unsuccessful attempt to realise this prediction. The existence of the Byzantine settlement Cosmidion before the Ottoman conquest, with several religious buildings in the area of today's Eyüp, is acknowledged, but described as 'not important' (Eyüpsultan Municipality Mayor's Office 1996, 6) and as having been destroyed by the crusades long before the conquest. Eyüp is portrayed as having played a significant role during the Ottomans' conquest of Istanbul in 1453. The books claim that, with the help of

Sheikh Akşemsettin, the grave of Ebu Eyyub was rediscovered there (Eyüp Belediyesi 2008b, 60; see also Coşkun 2015). With the conquest, Eyüp is portrayed as having found 'its real climate' and as having 'begun its function as a religious centre' (Eyüp Belediyesi 2008b, 67). Starting with the shrine and mosque complex commissioned by Mehmed II, the first Ottoman settlement outside Istanbul's city walls is said to have been founded as a completely 'new settlement' (Eyüp Belediyesi 2008a, 16), devoid of any influences from the Byzantine settlement, due to its previous destruction. Eyüp is portrayed as flourishing under Ottoman rule and called the 'most striking focal centre of Ottoman culture' (Eyüp Belediyesi 2008a, 16). The so-called 'Tulip Age', a period more generally referring to the first third of the eighteenth century (Erimtan 2008), is claimed to have constituted Eyüp's 'brightest period' (Eyüp Belediyesi 2008a, 16). 'Westernisation' is held responsible for the Golden Horn's industrialisation. Starting in the nineteenth century and increasing after the foundation of the Turkish Republic in 1923, industrialisation is blamed for having 'extinguished Eyüp's shining star' (Eyüp Belediyesi 2008a, 17). The narrative claims that Eyüp's environment 'started to grow like a tumour' with the settlement of Balkan migrants in the early republican years and migrants from Thrace and Anatolia from the 1960s (Eyüpsultan Municipality Mayor's Office 1996, 12). However, when the RP took office in 1994, it states that 'Eyüp entered a period of repair of the 150 years of erosion of this centuries-old historic identity' (Eyüp Belediyesi 2008b, 83).

This anti-migration discourse of the AKP is interesting, as many of their members have a personal migration background. However, declaring that migrants from regions in Europe, such as the Balkans and Thrace, started this process diverts the focus from the Anatolian migrants, who represent a large share of the AKP's supporters. Furthermore, the narrative primarily blames industrialisation related to 'westernisation' for the migrants' settling there, rather than the migrants themselves. The narrative also indicates that all Eyüp AKP mayors and many other AKP officials constitute or aspire to become part of the upper-middle class. In order to dissociate themselves from the poverty and lifestyles associated with the informal workers' settlements, they adapt an anti-migration rhetoric from other urban upper-middle class actors (Öncü 1999).

Eyüp as the stage of the Turkish War of Independence

The narrative of the foundation differs from the municipality's in several ways. The main elements of the narrative are illustrated in the exhibition of the foundation's museum, the Eyüp Historical Life and Culture Center. The exhibition was opened in one of the foundation's spaces in 2015 (Eyüp Dostları Vakfı n.d.-b).

Among old photos, documents, and everyday items from Eyüp's residents, a huge portrait of the republic's founding figure Mustafa Kemal Atatürk, flanked by a Turkish flag, is prominently visible.[5] The portrait was painted by Vesile Muzaffer Kargı (1925–2007). Her husband Mehmet Talip Kargı

(1924–1992) was the last sheikh of the Ümmi Sinan Lodge in Eyüp and grandson of Yahya Galip Kargı (1876–1942). According to the website of the association connected to the lodge, Yahya Galip Kargı was a strong supporter of the Turkish War of Independence. He became the governor of Ankara in 1919, and later a member of the Grand National Assembly. The website claims that Atatürk himself praised him for his services (Ümmisinan Türbesi Külliyesi Talip Kargı Türk Tasavvuf Müsikisi ve G. K. K. T. Y. Derneği n.d.; Türkmenoğlu 2005, 150–153). Next to the portrait, there is furniture from the Rıza Café, which was closed in 1991.[6] Another corner is filled with jerseys, trophies, and photos of the local football club, which according to Türkmenoğlu was engaged in underground activities during the War of Independence (2009, 36–38).[7]

The exhibition is based on documents, objects, and photos collected from people whom Türkmenoğlu calls '*eski Eyüplü*' (long-time Eyüp residents) and whom he defines as having lived in Eyüp for at least two to three generations (interview with author, 1 December 2016). Thus, in contrast to the municipality's narrative, the exhibition associates Eyüp's identity foremost with the social and cultural life of the last 100–150 years, which to a large part corresponds to the republican period. The foundation's narrative overlaps with the main arguments of the municipality's narrative, in the sense that both highlight Eyüp's spiritual significance, its importance for the Ottoman conquest, and Eyüp constituting a centre of Ottoman culture. Yet, the foundation changes the frame and adds elements of counter-narratives to the AKP's narrative. This becomes clear in the books edited by Türkmenoğlu.

The sections on Eyüp's history, written by Feridun Benden, attach a sacredness to Eyüp, as the municipality does, but not merely via the Prophet's Companion. Almost half of Benden's historical account in *Eyüp. Losing One's Heart to a Neighbourhood* dwells on Eyüp's pre-Ottoman history. It mentions not only Cosmidion, but also findings from Palaeolithic times, the antique Semystra altar, stories about the Golden Horn from Greek mythology, and a prediction by the Oracle of Delphi about the area's sacredness (2005). The wording of the oracle is interestingly similar to the hadith attributed to Muhammad about Istanbul's conquest. Benden writes: 'after some time passed, Muhammad heralded with almost the same words the sacredness of this town' (2005, 9). Although the contemporary place's sacredness is understood as being Islamic, a sacredness independent of Islam is also attributed to the place, and non-Muslims are permitted the authority to designate it as such.

Second, the foundation's narrative also stresses Eyüp as a centre of Ottoman culture. Yet, the centrality does not arise primarily from religious scholars and institutions, or Ottoman architectural structures as the municipality claims, but from the presence of the Ottoman state elite living in Eyüp. In a video from the television programme *Kayda Değer*, Türkmenoğlu describes life in Eyüp as having been thriving with 'culture' in the past due to the musicians, poets, and thinkers among this distinguished community. According to him, people were permitted to settle in Eyüp during Ottoman times only after the local governor (*kadı*) decided the applicant was 'fitting' and 'qualified'. He characterises Eyüp's residents as having constituted a group of 'upper level bureaucrats, officers and statesmen' until the 1950s. He contrasts this imagined distinctive and 'cultured' social composition of the past with the migrants coming after the 1950s, whom he describes as 'workers' who completely changed the social fabric (Eyüp Dostları Vakfı n.d.-c).

Furthermore, while the municipality portrays Eyüp as Ottoman heritage, Türkmenoğlu's books often frame it as Turkish heritage. In fact, Eyüp is presented not only as Turkish heritage, but as the heritage of humankind on the same level with ancient Egypt, Rome, or Delphi (Benden 2005, 9). Linking Eyüp to classical antiquity serves to establish a connection between it and the imagined community it represents with an imagined 'western civilisation', as republican or Kemalist narratives tend to do (Fisher Onar 2015, 149; see also Shaw 2007, 258). The members of the foundation view the foundation of the republic and its founding figure Mustafa Kemal Atatürk very positively. In the exhibition, apart from the portrait, several objects suppose a connection between Eyüp's residents and Atatürk. The foundation's members stress the support of Eyüp's residents for the War of Independence, which laid the ground for the foundation of the republic. In his book on the Eyüp sports club, Türkmenoğlu emphasises the club's importance during this war. He describes in detail the active involvement of individuals and institutions in Eyüp in the nationalist resistance movement. The sports club, he argues, transported weapons to Istanbul's Anatolian side via activities camouflaged as sports competitions. The Reşadiye Numune Mektebi school and Hatuniye Lodge are described as places for weapon storage (2009, 20–22, 36–41). Many characters in the book, such as Yahya Galip Kargı of the Ümmi Sinan Lodge's family and the sheikh of Hatuniye Lodge, are portrayed as having a strong Muslim religious identity and as supporting the War of Independence at the same time.

The periodisation of the historical narrative is as follows: Eyüp's settlement history goes back to the Palaeolithic period, with a pilgrimage site in the area since antiquity and Byzantine times at the latest. In contrast to the municipality's narrative, Benden writes that Cosmidion's Byzantine church 'continued its sacredness almost until the [Ottoman] conquest of Istanbul' as a Christian pilgrimage place (2005, 10). Thanks to their military tactics, which are described in detail, the Turks manage to conquer Istanbul and realise the prophecy about the conquest in contrast to the Arabs, whose repeated failures to do so are mentioned several times. The discovery of Ebu Eyyub's grave, which in the municipality's narrative is always related to the conquest, is not even mentioned in *Eyüp. Losing One's Heart to a Neighbourhood*. It only notes Sheikh Akşemsettin's prediction about the conquest and his support as a necessary 'spiritual bullet' (Benden 2005, 10–12). The Turks are described as more successful than not only the Arabs, but also than many other 'foreign peoples and nations', who had tried to conquer Istanbul due to its geopolitical, economic, and cultural importance (Benden 2007, 4).

As in the municipality's narrative, Eyüp is said to have flourished during Ottoman times. The Turkish War of Independence signifying the end of the Ottoman Empire and resulting in the establishment of the Turkish Republic is as glorious for the Turks as the Ottoman conquest of Istanbul, and Eyüp's residents are portrayed as having played an important part in supporting this war. The early republican times are narrated as a period when Eyüp's social fabric was still intact, while Türkmenoğlu describes the time between the 1930s and 1950s as a 'turning point', when huge numbers of migrants settled in Eyüp, often informally, and when the area started to be polluted by industrial expansion in the Golden Horn. This resulted in many 'old' Eyüp people leaving the neighbourhood (interview with author, 1 December 2016). This narrative has a 'golden age' and a period of decline, but it does not include a period of revival.[8]

Monuments as living history and documentation of a bygone past

In addition to the content of the narratives, the two groups also differ in how they engage with their narratives and how they authenticate and communicate them. The municipality portrays the destruction and degradation of physical structures as a decline of Eyüp's identity, and their renovation and repair as revival. As such, the municipality claims to have revived Eyüp's identity. Many residents remember that until recently, Eyüp's environment not only looked but also smelt very differently due to the Golden Horn's industrial pollution. On the one hand, this makes it possible to construct a period of decline and portray the RP government after 1994 as the saviour of Eyüp's neglected identity. On the other hand, the recent transformations of Eyüp's physical fabric also point to the fact that what the municipality presents as Eyüp's 'original' fabric is actually a relatively recent construction. In the first years after 1994, the new government strongly advertised its physical interventions, which it called 'restoration' activities. These also included the erection of structures without direct historical examples, but the style of which was presented as 'fitting to the original' (Eyüp Belediyesi 2008a, 202; see also Hammond 2016, 98–100). Up to the 2000s, the municipal guides called Eyüp an 'open air museum' (Eyüp Belediyesi 2008b, 67), marking Eyüp as an important remnant of the past that needed to be preserved. Recently, after major interventions in the physical fabric, an unbroken continuity with the past has been suggested. Thus, in contrast, the 2016 guidebook claims Eyüp to be 'not a museum […] but a piece of living history' (Aydın 2016).

Given the importance of physical locations and landscape for the construction of identity narratives, their design can be an important tool for disseminating such narratives. Materialising a particular identity narrative in the physical landscape, along with markers such as monuments, both requires and demonstrates political power (Hodgkin and Radstone 2003, 11–13; Mills 2011). A physical environment not only makes a narrative visible, but it also powerfully shapes the way people move through and experience the place.

This certainly affects the everyday lives of Eyüp's residents. Moreover, the transformation of Eyüp's fabric by governmental actors is part of branding Eyüp with a particular Islamic-Ottoman image for tourists and potential future residents (Roelcke 2019). The municipality tries to shape the interpretation and use of physical structures in the sense of creating *lieux de mémoire*, sites where particular forms of memory related to the AKP's identity narratives are promoted (Walton 2016). The structures used to present this narrative, mainly Ottoman-era monumental architecture, can easily be understood by diverse audiences as representing the AKP's narratives about a glorious Islamic-Ottoman identity. But as many of the sites, such as mosques and adjacent squares, are open to the public, they are appropriated in diverse ways by various people including residents and visitors, and the municipality's framing of the places is not uncontested (Hammond 2016, 145–188; Nikro 2019).

The members of the Foundation of Eyüp's Friends do not currently have the political power to shape the physical landscape according to their narrative. Their nostalgia for an earlier version of Eyüp corresponds to times when socio-political constellations were different, before the workers settled there since the 1950s and before the RP took office.

They do not aim to recreate the neighbourhood based on their nostalgia of Eyüp. Instead, assuming that Eyüp's original identity and its physical remnants are vanishing, they try to collect documents on Eyüp's history based on their historical narrative, focusing on the first half of the twentieth century. The municipality presents contemporary and coloured photographs in its books to illustrate the current revival of Eyüp's imagined identity. Türkmenoğlu's books instead use photographs from the beginning of the twentieth century to represent the 'times of harmony' in Eyüp. The old black-and-white photos serve to document the imagined original as a point of reference, to which the present can only look different.

Türkmenoğlu expresses a motivation to 'record' 'the last representatives of an epoch and a culture in Eyüp', including through his oral history book *Eyüp's Story of the Last Century—From the Speech of the Living Ones*. The book resulted from interviews with 120 people aged over 60 (Türkmenoğlu 2018, 6). Having become a more widespread phenomenon globally in recent years, and often conducted outside the academic context, oral history is frequently used to present alternative historical narratives. Frequently, individuals' memories are attributed authenticity and truth capable of challenging established narratives (Hodgkin and Radstone 2003, 4). Likewise, Türkmenoğlu claims to present less well-known aspects of the neighbourhood, 'next to Eyüp's sacred structure also its modern and innovative sides' (Türkmenoğlu 2018, 7). By referring to 'witness accounts', Türkmenoğlu authenticates his narratives through a selectively chosen group of people, whom he calls *eski Eyüplü* (long-time Eyüp residents) or *köklü Eyüplü* (rooted Eyüp residents). This group constitutes the core of all his projects. Even if several of them have moved away from Eyüp, he still considers them to be *Eyüplü*. According to Türkmenoğlu, these people, who 'experienced […] and shaped the last century', are the 'main source [for the oral history, book, and museum projects], because they have everything [information, photographs, objects …]. The people of today's Eyüp have nothing [like this]' (interview with author, 1 December 2016). This clearly indicates that the past to be remembered, in his opinion, is the time before the majority of today's Eyüp's residents arrived, and if later, then activities which they were not part of. Talking about the residents of today's Eyüp, Türkmenoğlu uses the term '*Eyüp'tekiler*', those who are in Eyüp, and not '*Eyüplüler*', those belonging to Eyüp, as if they were, even after decades, still foreign to the place.

The purpose of documenting Eyüp's past is to keep alive what the foundation's members call a 'true Eyüp spirit' (Türkmenoğlu 2009, 7) through remembering. İlhan S. Ergelen, the publisher of *A Moment Comes When Time Stops … Eyüp*, claims to 'relive' his childhood when looking at the photos presented in his book. This is interesting, as Ergelen was, according to his preface, born in the 1940s (2007, 3). However, most of the photos in the book were taken before he was born, most of them during Ottoman times. If Ergelen feels able to relive his childhood based on these photos, this suggests that in his conception of Eyüp's history the neighbourhood did not change much between the late nineteenth century until his childhood in the 1940s, and that the major changes constituting a substantive difference between contemporary times and those times happened only later, such as with the arrival of migrants in the 1950s. Ergelen's statement may also point to the importance attributed to family lineage, as he may conceive himself as being the heir to his parents' and grandparents' memories (Hodgkin and Radstone 2003, 10).

The museum does not have any signs on the building or in the streets leading to it, and so is only visited by people who already know about it. The foundation does reach out among residents, for example at local schools. But generally, and in contrast to the municipality, the foundation's members prefer to address those they perceive as already belonging to their social group. Their activities serve to strengthen social cohesion within a certain group of formerly dominant middle and upper-middle class people generally perceived as secular. They feel marginalised by the shifting social and power constellations. Their activities also serve to assert difference to currently dominant identity narratives about Eyüp and its residents. Rather than informing outsiders about Eyüp's history, their museum primarily functions as a meeting place for the foundation's members. In addition to conveying the actual content of historical narratives, both the processes of creating and engaging with the *lieux de mémoire*, such as the museum or the oral history archive, stimulate social exchange and thus can be understood as creating *milieux de mémoire*, environments of memory (Nikro 2019, 11).

Conclusion

The chapter has demonstrated the different ways in which the Foundation of Eyüp's Friends and the AKP-ruled Municipality of Eyüp narrate Eyüp's identity and history, and in which they authenticate and communicate their narratives within contemporary power struggles and shifting social, economic, and political constellations.

The AKP-ruled Municipality of Eyüp narrates Eyüp as the centre of an imagined Islamic-Ottoman civilisation in the context of the AKP's identity politics locally and on a wider level. It portrays the Ottoman era as Eyüp's 'golden age'. The Ottoman Empire's last years and republican times feature as a period of decline, while the municipality claims to have revived Eyüp's imagined original identity since the RP took office in 1994. The municipality projects Eyüp's identity onto physical structures and presents their restoration as identity revival. The structures can be easily understood as symbolising the AKP's general narratives by diverse audiences they target.

The foundation's members feel marginalised due to the rise of the AKP and its predecessors. Trying to assert difference to the currently dominant social groups, the foundation's members insist on an originally elitist character of Eyüp during Ottoman and early republican times. They claim Eyüp's 'original' identity was lost with the arrival of workers starting in the 1950s, whom they present as culturally 'other'. By documenting Eyüp's pre-1950 past based on the accounts of witnesses, they aim to challenge the currently dominant narrative.

The sites of memory created by the municipality provide the opportunity for social exchange to diverse people and multiple ways of appropriation via their accessibility and presence in public space. On the other hand, in the foundation's case the very production of sites of memory and their functioning as a meeting point serve to strengthen feelings of belonging to a social group that positions itself as different to the dominant one.

While the two narratives appear to contrast by asserting a difference between the groups their narrators represent, the structures of their accounts exhibit many similarities. Both establish periodisations of the neighbourhood's history, albeit in different ways, based on the assumption of the existence of an 'original' identity of Eyüp. They claim it was present in a glorious past but was then lost during a period of decline, with nostalgic longing left in its wake. Both portray Eyüp as a stage for decisive turning points in the development of their imagined larger communities, with male heroes featuring as actors bringing about the change.

Furthermore, those residents who settled in Eyüp after the 1950s are subsumed under the period of decline in both narratives, portrayed as destroying Eyüp's physical fabric and changing its culture. This resembles a general narrative among the urban middle and upper-middle classes that portrays rural migrants to the large cities as the cultural 'Other' (Öncü 1999). The municipality locates these migrants' origins in Anatolia, as well as in former Ottoman areas and European Turkey, whereas the foundation's members locate their origin mainly in Anatolia. In opposition to them, they connect Eyüp to an imagined modern and 'western' civilisation. Linked to that, both narratives mainly project Eyüp's imagined identity onto a relatively small area in the administrative district's historical centre. They generally disregard most of the district, where the majority of those coming after the 1950s settled.

Both the municipality and the foundation use frames of religious and cultural significance in their narratives, but they do so in different ways. In the narratives of both, Eyüp's Ottoman past features prominently in the construction of the neighbourhood's 'true' identity. However, Türkmenoğlu's books frame Eyüp's Ottoman heritage and its Islamic significance as part of the heritage of humankind, connecting it to a western heritage discourse, while the municipality frames Eyüp as the heritage of Islam, conceiving Islam as a civilisation.

The narrative of the foundation's members constitutes a Kemalist narrative in their focus on the Turkish War of Independence, their attempts to prove support and connections between Eyüp's residents and Atatürk, and their stressing of a Turkish identity. Their claim that 'secular republican' values are threatened by rural migrants from Anatolia and by Islamist movements—the AKP often being perceived as representing both—are further elements of this discourse (Bora 2003; Fisher Onar 2015; Çınar and Taş 2017).

At the same time, the foundation's members' narrative embraces a common discourse about Eyüp as being an Islamic place, and as being important to Istanbul's Ottoman conquest and heritage. Thus, the foundation's members regard Ottoman times as a main element of Eyüp's 'golden age', with the shrine of Ebu Eyyub, seen as an Islamic saint, as a significant element of the neighbourhood's identity. The figures connecting Eyüp with the War of Independence and Atatürk have a clear Muslim background. Rather than distinguishing between secular and religious identities, the members of the Foundation of Eyüp's Friends contrast the urban and the rural, different social classes, and residents' European and Anatolian backgrounds. Thereby, their narrative complicates the perceived dichotomy of the religious and the secular. The particularity of the foundation's narrative about Eyüp illustrates the uniqueness of socially constructed places as particular constellations of connections beyond the places' imagined boundaries (Massey 1995).

Notes

1 On guidebooks and identity construction, see Koshar (1998).
2 (denizkoken.com, n.d.; liderler.net, n.d.; yerelhaberim.net, n.d.).
3 (habereyup.com, 2014; eflatunhaber.com, 2018).
4 For changes in the Municipality of Eyüp's narrative on Eyüp with changing socio-political contexts based on an analysis of the guidebooks see Roelcke (2019).
5 The image can be accessed on the Foundation of Eyüp's Friends' website: www.edv.org.tr/wp-content/uploads/2017/05/19-1.jpg.
6 The image can be accessed here: www.edv.org.tr/wp-content/uploads/2017/05/17-1.jpg.
7 The image can be accessed here: www.edv.org.tr/wp-content/uploads/2017/05/12-1.jpg.

8   The narrative of the Foundation of Eyüp's Friends resembles narratives of urban planners who published about Eyüp, and who also belong to the urban upper-middle class
(Roelcke 2020).

References

Acarer, Erk. 2017. 'Erdoğan'ın Eyüp Sultan'da kıldığı şükür namazının anlamı: 'Saltana-
tımı ilan ettim' mesajı.' *Birgün.net*. 17 April 2017. www.birgun.net/haber/erdogan-in-
eyup-sultan-da-kildigi-sukur-namazinin-anlami-saltanatimi-ilan-ettim-mesaji-155852
Accessed 6 January 2020.

Anderson, Benedict. 1991. *Imagined Communities: Reflections on the Origin and Spread
of Nationalism*. London: Verso.

Artan, Tülay. 1998. 'Terekeler Işığında 18. Yüzyıl Ortasında Eyüp'te Yaşam Tarzı ve Stan-
dartlarına Bir Bakış – Orta Halliliğin Aynası.' In *18. Yüzyıl Kadı Sicilleri Işığında
Eyüp'te Sosyal Yaşam*, edited by Tülay Artan, 49–64. Istanbul: Tarih Vakfı Yurt
Yayınları.

Aydın, Remzi. 2016. 'Eyüp'e Hoşgeldiniz.' In *Eyüp Gezi Rehberi*, edited by Kutse Özafşar
and Adem Uyar, 1. Istanbul: Eyüp Belediyesi.

Bartu, Ayfer. 1999. 'Who Owns the Old Quarters? Rewriting Histories in a Global Era.' In
Istanbul: Between the Global and the Local, edited by Çağlar Keyder, 31–45. Boulder,
New York, and Oxford: Rowman & Littlefield.

Benden, Feridun. 2005. 'Tarihçe.' In *Eyüp. Bir Semte Gönül Vermek*, edited by
Şener Türkmenoğlu, 9–12. Istanbul: ABC Yayın Grubu.

Benden, Feridun. 2007. 'Tarihçe.' In *Bir An Gelir, Durur Zaman ... Eyüp*, edited by
Şener Türkmenoğlu, 4–6. Istanbul: Hazinedaroğlu İnşaat Grubu.

Bezmez, Dikmen. 2008. 'The Politics of Urban Waterfront Regeneration. The Case of
Haliç (the Golden Horn), Istanbul.' *International Journal of Urban and Regional
Research* 32 (4): 815–840.

Bora, Tanıl. 2003. 'Nationalist Discourses in Turkey.' *South Atlantic Quarterly* 102 (2/3):
433–451.

Çakır, Fedai. 2016. 'Eyüp Sultan Aşk'ı.' 17 January 2016. http://fedaicakir.com/tag/ismail-
kavuncu/.

Chauvel, Brian. 2011. 'Retour' et 'reconquête' de la péninsule historique: discours
et usages distinctifs autour du patrimoine de Fener et Çarşamba.' *EchoGéo*
16: 1–28.

Christophe, Barbara, Christoph Kohl, and Heike Liebau. 2017. 'Politische Dimensionen
historischer Authentizität: Lokale Geschichte(n), (Macht-)Politik und die Suche nach
Identität.' In *Geschichte als Ressource. Politische Dimensionen historischer Authentizität*, edited by Barbara Christophe, Christoph Kohl, and Heike Liebau, 9–33. Berlin:
Klaus Schwarz Verlag.

Çınar, Alev. 2001. 'National History as a Contested Site: The Conquest of Istanbul and
Islamist Negotiations of the Nation.' *Comparative Studies in Society and History* 43 (2):
364–391.

Çınar, Alev, and Hakan Taş. 2017. 'Politics of Nationhood and the Displacement of the
Founding Moment: Contending Histories of the Turkish Nation.' *Comparative Studies
in Society and History* 59 (3): 657–689.

Coşkun, Feray. 2015. 'Sanctifying Ottoman Istanbul: The Shrine of Abū Ayyūb al-Anṣārī.'
PhD diss., Freie Universität Berlin.

denizkoken.com. n.d. 'Biyografisi.' Accessed April 25, 2019.www.denizkoken.com/default.asp?sayfa=hakkinda&bolum=biyografi.

eflatunhaber.com. 2018. 'İyi Parti Eyüp Belediyesi'ne talip!' March 7, 2018. www.eflatun haber.com/eyup-sultan/iyi-parti-eyup-belediyesi-ne-talip-h29400.html.

Ergelen, İlhan S. 2007. 'Sunum.' In *Bir An Gelir, Durur Zaman ... Eyüp*, edited by Şener Türkmenoğlu, 3. Istanbul: Hazinedaroğlu İnşaat Grubu.

Erimtan, Can. 2008. *Ottomans Looking West? The Origins of the Tulip Age and Its Development in Modern Turkey*. London: Tauris Academic Studies.

Eyüp Belediyesi. 2008a. *Eyüp'ün Altın Yılları '94-'09*. Istanbul: Eyüp Belediyesi.

Eyüp Belediyesi. 2008b. *Eyüp Şehir Rehberi. The City Guide*. Istanbul: Eyüp Belediyesi.

Eyüp Dostları Vakfı. 2017. 'Eyüp Dostları Vakfı Senedi.' 14 March 2017. https://edv2017.word press.com/2017/03/14/eyup-dostlari-vakfi-senedi/.

Eyüp Dostları Vakfı. n.d.-a. 'Eyüp Dostları Vakfı Kuruluş Temeli.' Accessed 14 April 2018. www.edv.org.tr/tarihce/.

Eyüp Dostları Vakfı. n.d.-b. 'Hikayemiz.' Accessed 14 April 2018. www.edv.org.tr/muze mizin-baslangic-hikayesi/.

Eyüp Dostları Vakfı. n.d.-c. 'EDV Kayda Değer.' Basında Tarihi Yaşam ve Kültür Merkezimiz, Video. Accessed 14 April 2018. www.edv.org.tr/basinda-tarihi-yasam-ve-kultur-merkezimiz/.

Eyüpsultan Municipality. n.d. 'Başkan – Özgeçmiş.' Accessed 12 April 2018. www.eyup sultan.bel.tr/tr/main/pages/ozgecmis/9.

Eyüpsultan Municipality Mayor's Office. 1996. *Eyüp Sultan Rehberi 1996*. Istanbul: İki Nokta.

Faroqhi, Suraiya. 1998. 'Migration into Eighteenth-Century 'Greater Istanbul' as Reflected in the Kadı Registers of Eyüp.' *Turcica* 30: 163–183.

Fisher Onar, Nora. 2015. 'Between Memory, History, and Historiography: Contesting Ottoman Legacies in Turkey, 1923–2012.' In *Echoes of Empire: Memory, Identity and Colonial Legacies*, edited by Kalypso Nicolaidis, Berny Sebe and Gabrielle Maas, 141–154. London and New York: I.B. Tauris.

habereyup.com. 2014. 'Ferzan Özer: Eyüp'ü yönetecek bilgiye, birikime, heyecana sahibiz.' 5 March 2014. www.habereyup.com/ferzan-ozer-eyupu-yonetecek-bilgiye-biri kime-heyecana-sahibiz_4249.html.

Hammond, Timur Warner. 2016. 'Mediums of Belief: Muslim Place Making in 20th Century Turkey.' PhD diss., University of California, Los Angeles.

Hodgkin, Katharine, and Susannah Radstone. 2003. 'Introduction: Contested Pasts.' In *Contested Pasts: The Politics of Memory*, edited by Katharine Hodgkin and Susannah Radstone, 1–21. New York and London: Routledge.

Huyssen, Andreas. 2003. *Present Pasts. Urban Palimpsests and the Politics of Memory*. Stanford, CA: Stanford University Press.

Karaosmanoglu, Defne. 2010. 'Nostalgia Spaces of Consumption and Heterotopia: Ramadan Festivities in Istanbul.' *Culture Unbound: Journal of Current Cultural Research* 2 (2): 283–302.

Koshar, Rudy. 1998. "What Ought to Be Seen': Tourists' Guidebooks and National Identities in Modern Germany and Europe.' *Journal of Contemporary History* 33 (3): 323–340.

Kresse, Kai. 2017. 'Thinking with Internal Comparisons: 'Implicit Comparison' and 'Mutual Perceptions'.' Presentation at the symposium Modalities of Co-existence Across Religious Difference: Critical Terms for the Study of Indigenous Religion,

Christianity, and Islam in Plural Settings in Africa and Beyond. Leibniz-Zentrum Moderner Orient, Berlin, 22–24 March.

liderler.net. n.d. 'Eyüp Belediye Başkanı Sn. Ahmet Genç.' Accessed 12 April 2018. www .liderler.net/haberler/827/eyup_belediye_baskani_sn_ahmet_genc.html.

Massey, Doreen. 1995. 'Places and Their Pasts.' *History Workshop Journal* 39: 182–192.

Mills, Amy. 2005. 'Narratives in City Landscapes: Cultural Identity in Istanbul.' *Geographical Review* 95 (3): 441–462.

Mills, Amy. 2011. 'The Ottoman Legacy: Urban Geographies, National Imaginaries, and Global Discourses of Tolerance.' *Comparative Studies of South Asia, Africa and the Middle East* 31 (1): 183–195.

Nikro, Norman Saadi. 2019. *Milieus of ReMemory. Relationalities of Violence, Trauma, and Voice.* Newcastle upon Tyne: Cambridge Scholars Publishing.

Nora, Pierre. 1989. 'Between Memory and History: Les Lieux de Mémoire.' *Representations* 26: 7–24.

Öncü, Ayşe. 1999. 'Istanbulites and Others. The Cultural Cosmology of Being Middle Class in the Era of Globalism.' In *Istanbul: Between the Global and the Local*, edited by Çağlar Keyder, 95–119. Lanham, Boulder, New York, and Oxford : Rowman & Littlefield.

Öncü, Ayşe. 2007. 'The Politics of Istanbul's Ottoman Heritage in the Era of Globalism. Refractions Through the Prism of a Theme Park.' In *Cities of the South. Citizenship and Exclusion in the Twenty-First Century*, edited by Barbara Drieskens, Franck Mermier and Heiko Wimmen, 233–264. London: Saqi Books.

Öztürk, Ayşe. 2017. 'Zwischen Spiritualität und Kulturtourismus: Neo-Osmanische Stadtentwicklung in Eyüp, Istanbul.' *Inamo* 89: 39–42.

Özyürek, Esra. 2006. *Nostalgia for the Modern: State Secularism and Everday Politics in Turkey.* Durham, NC: Duke University Press.

Roelcke, Annegret. 2019. 'Constructing the Capital of Peace: Changing Branding Strategies for Istanbul's Eyüp Quarter.' *Middle East – Topics & Arguments* 12: 110–120.

Roelcke, Annegret. 2020. 'Pre-AKP Urban Rehabilitation Projects for Istanbul's Eyüp Quarter: Contextualising the Narrative of 1994 as Point of Rupture.' In *Türkeiforschung im Deutschsprachigen Raum. Umbrüche, Krisen und Widerstände*, edited by Johanna Chovanec, Gabriele Cloeters, Onur Inal, Charlotte Joppien and Urszula Wozniak, 207–229. Wiesbaden: Springer VS.

Shaw, Wendy. 2007. 'Museums and Narratives of Display from the Late Ottoman Empire to the Turkish Republic.' *Muqarnas* 24: 253–279.

Trouillot, Michel-Rolph. 2015. *Silencing the Past. Power and the Production of History.* Boston: Beacon Press.

Tuğal, Cihan. 2009. 'The Urban Dynamism of Islamic Hegemony: Absorbing Squatter Creativity in Istanbul.' *Comparative Studies of South Asia, Africa and the Middle East* 29 (3): 423–437.

Türkmenoğlu, Şener. 2005. *Eyüp. Bir Semte Gönül Vermek.* Istanbul: ABC Yayın Grubu.

Türkmenoğlu, Şener, ed. 2007. *Bir An Gelir, Durur Zaman ... Eyüp.* Istanbul: Hazinedaroğlu İnşaat Grubu.

Türkmenoğlu, Şener. 2009. *90. Yılında Eyüp Spor Kulübü ve İlçe Kulüpleri Tarihi.* Istanbul: ABC Yayın Grubu.

Türkmenoğlu, Şener, ed. 2018. *Son Yüzyılın Hikayesi Eyüp -Yaşayanların Dilinden.* Istanbul: Yayın Dünyamız Yayınları.

Ümmisinan Türbesi Külliyesi Talip Kargı Türk Tasavvuf Müsikisi ve G. K. K. T. Y. Derneği. n.d. 'Yahya Galip Kargı.' Accessed 05 July 2019. www.ummisinan.org/yahya-galip-karg.

Walton, Jeremy. 2016. 'Geographies of Revival and Erasure: Neo-Ottoman Sites of Memory in Istanbul, Thessaloniki, and Budapest.' *Die Welt des Islams* 56 (3–4): 511–533.

Winter, Jay M. 2009. 'In Conclusion: Palimpsests.' In *Memory, History, and Colonialism: Engaging with Pierre Nora in Colonial and Postcolonial Contexts*, edited by Indra Sengupta, 167–173. London: German Historical Institute London.

Yaramış, Ahmet. 2004. 'Feshane'nin İlk Kuruluş Yılları.' In *Tarihi, Kültürü ve Sanatıyla Eyüpsultan Sempozyumu VIII Tebliğler*; [7–9 Mayıs 2004], edited by Municipality of Eyüp, 94–99. Istanbul: Eyüp Belediyesi Kültür ve Turizm Müdürlüğü.

Yenen, Zekiye, Akın, Oya, and Yakar, Hülya, eds. 2000. *Eyüp: Dönüşüm Sürecinde Sosyal- Ekonomik-Mekansal Yapı*. Istanbul: Eyüp Belediyesi Yayınları.

yerelhaberim.net. n.d. 'İsmail Kavuncu Kimdir?' Accessed 12 April 2018. www.yerelha berim.net/ismail-kavuncu–kimdir-biyografi,28.html.

5 Past neighbourhoods

Palestinians and Jerusalem's 'enlarged Jewish Quarter'

Johannes Becker

Introduction

This chapter deals with neighbourhoods that have ceased to exist, and with questions such as what it means to have been forced to leave them and to have only fading memories of them, or what it means to have remained in the area where they once were. The area of Jerusalem's 'enlarged Jewish Quarter' – a term borrowed from Michael Dumper (2002, 14) – was administratively defined as ethno-religiously exclusive by the Israeli authorities after the conquest of the Old City in 1967. Its definition and borders were considerably widened in comparison to the shifting area previously identified as the Jewish Quarter, and it encompasses several former, ethno-religiously diverse neighbourhoods. The Palestinian residents of this area were nearly all evicted between 1967 and 1977 within the framework of the redefinition of space called the 'reconstruction' of the Jewish Quarter. A small minority of Palestinians was allowed to remain there.

The enlarged Jewish Quarter is one of three spaces in Jerusalem's Old City within which I have conducted research on emplacement as a spatial and temporal process (Becker, 2016, 2017a).[1] This chapter is based on eight biographical-narrative interviews (Schütze, 1983, 2014; Rosenthal, 2004) with Palestinians whose families remained living as outsiders in the enlarged Jewish Quarter, and four interviews with Palestinians who were evicted in the late 1960s or the 1970s.

On the basis of case reconstructions of one former resident and one current resident of the enlarged Jewish Quarter, I argue that the former Palestinian neighbourhoods have been placed in the past, not only as a result of the evictions, but also due to a process of marginalisation on two levels: the restriction of everyday life for those who remained there (*spatial marginalisation*), and the waning memory of these neighbourhoods (*temporal marginalisation*). First, I ask to what extent the former Palestinian neighbourhoods, and individual memories of life there, are undergoing a process of *temporal marginalisation* and are not part of Palestinian collective memory. To answer this question, I reconstruct the *memories of place* expressed in the self-presentation of a former neighbourhood resident within a biographical interview, as well as in his interactions within his family. Second, I reconstruct the perceptions of places and connected experiences as *processes of emplacement* of those who continue to live there. The everyday living conditions determined by the figuration of Palestinians and Jewish Israelis, as well as the Israeli top-down rebranding of space, prohibit the continuation of previous neighbourhood life and *spatially* marginalise the neighbourhoods.

For these reasons, the neighbourhoods in the extended Jewish Quarter have lost qualities otherwise ascribed to them in other Old City areas. Palestinian neighbourhoods there continue to symbolise security and social control; they are a (positive or negative) reference point and present a possible we-image for their inhabitants. Neighbourhood belonging usually offers mutual identification and assurance and can be used as a self-ascription of emplacement. Although all Palestinian Old City neighbourhoods are under pressure by expansionist politics, the ones in the enlarged Jewish Quarter have practically vanished – in terms of everyday life and increasingly fading memories. While Israeli policies are generally a driving force behind this marginalisation, it is also connected to intra-Palestinian power dynamics between long-established and internal migrants, as well as between men and women.

The case of Jamal, a former resident whose family was evicted from one of the neighbourhoods (the *Ḥārat aš-Šaraf*) highlights the extent to which the Israeli definition of the enlarged Jewish Quarter has been accepted, even by Palestinians who were evicted from there, and that the fading memory of the neighbourhood is framed as one of a dirty past. The former inhabitants lack a common language to talk about their expulsion from these neighbourhoods. Therefore, the former neighbourhoods that were incorporated into the enlarged Jewish Quarter seem to be in the process of being excluded from Palestinian collective memory. This highlights a successful framing of these spaces as exclusively Jewish by Israeli spatial politics, but it is also connected to the fact that many of the Palestinians who left the neighbourhoods were refugees or internal migrants with a weaker power position within Palestinian Jerusalemite society.

The case of Huda, a young woman in the enlarged Jewish Quarter, highlights how the hostile living conditions and lack of neighbourhood life for Palestinians who still live there, intertwined with familial control, lead to her forced confinement in the family house. Huda transforms this position by framing her biography as part of the historical and political significance of Jerusalem as an alternative emplacement. Whereas emplacement in the neighbourhood is still an important communalising reference point in other Palestinian areas in the Old City, I observed that isolation within the enlarged Jewish Quarter has often led to a more individualising position of the inhabitants there.[2]

As a way of framing the discussion on 'past neighbourhoods', I will briefly introduce the relationship between biographies and place: first, the relation between biographical courses and places as *biographical emplacement* – as a temporal and spatial process – and, second, memories of place during self-presentations in the interviews.

Places and biographies

In addition to being religiously and politically charged, the walled Old City is also a very crowded and precarious living space. The whole area comprises less than one square kilometre, of which only 0.6 square kilometres are used as residential neighbourhoods. More than 40,000 people live in this confined space, including Jews, Christian clerics from very different parts of the world, and members of different Palestinian groupings who form the large majority, and who are often at the lower end of the socio-economic scale. To understand the status of the Old City in past decades, it is important to consider the cramped, socially and economically difficult living conditions, the large-scale internal migration to the Old City in the past one hundred years, and the devaluation of the Old City in the intra-Palestinian discourse. Apart from the city's religious and political significance, it is probably for these reasons that space is negotiated more explicitly than in other local contexts.

In the interviews I conducted in Jerusalem, not only was the Old City as a whole an important reference point, but so were certain places *within* the Old City, such as certain neighbourhoods, buildings, or symbolic images of Jerusalem. These are connected to individual and familial experiences in the Old City, to the history of groupings living there, but also to discourses on the Old City, collective memories, and the current political situation. Neighbourhoods were especially important in both narrations and reconstructions of everyday life. Even if they play a less important role today than they did previously, they still contribute to the city's spatial and social organisation and provide orientation within life stories. The seemingly individual construction of places, such as neighbourhoods, in life stories also tells us something about the spatial processes of constructing belonging: '[The] construction of places, in the sense of known and definable areas, is a key way in which groups and collectivities create a shared, particular and distinctive identity' (McDowell, 1997, 2). All the more significant was my finding that both former and current Palestinian residents of the enlarged Jewish Quarter accepted the Israeli-imposed definition of that space. In the life-story interviews I conducted, there was no apparent connection with the former Palestinian neighbourhoods. This is what I will explore in this chapter.

Places have largely been excluded from research based on biographies and family histories. Sociological biographical research looks at how life stories are both socially constructed and individually shaped at the same time. Most researchers, at least in the German-speaking context, work with a *heuristic* separation of life history and life story. This means attempting to reconstruct biographical courses and how experiences are interrelated with social history (*life history*), on the one hand, and reconstructing self-presentations in the present of the interview situation (*life story*), on the other (see for example Rosenthal, 2004). This heuristic separation serves as a basis for my analyses in this chapter: I reconstruct perceptions of places and connected experiences, or *biographical emplacements*, as a temporal and spatial process in the life history of Huda, who still lives in the enlarged Jewish Quarter. I also reconstruct *memories of place* in the context of life stories, as in the case of Jamal, a former resident of one of the Palestinian neighbourhoods. Using these two case reconstructions, I analyse the process that has led to these neighbourhoods being confined to the past. I argue that this was not only due to the evictions, but that the process of marginalisation continued on two levels: *spatial marginalisation*, as a way of restricting those who stayed there, and *temporal marginalisation* in the sense of the waning memory of these neighbourhoods. Especially in terms of the latter, these former neighbourhoods differ from the former villages of pre-1948 Palestine, which play a central role in Palestinian collective memory.

The basic premise when speaking about places and life stories is the assumption that places and individuals are mutually constitutive. Linda McDowell (1997, 1, cf. Massey, 1995) argues that 'there is a reciprocal relationship between the constitution of places and people. Thus, there is a dual focus on how places are given meaning and how people are constituted through place.' This idea of the reciprocal constitution of places and people widens the often-held assumption that places are formed by mobile and immobile things, and forms the basis of a connection between places and life histories or life stories. Places are thus seen as including social interactions, as well as memories, myths, and discourses connected with them (Massey, 1994, 155; McDowell, 1997, 2). If embedded in a socio-historical or sociological analysis, the reconstruction of life stories is a suitable tool to prevent the essentialist assumption that everyone experiences and remembers places in the same way. Taking the idea of the co-constitution of places by people seriously, the reconstruction of life stories allows us to say something not only about individual experiences, but also about the places that they co-constitute. This may help contribute to overcoming the often-criticised separation of assumed objective social reality from subjective perceptions of it (Apitzsch and Inowlocki, 2000, 53–58).

The interconnected analysis of places and biographical experiences leads to the study of *biographical emplacement* as a temporal and spatial process (Becker, 2017a). Analysing emplacement considers the importance of primary and secondary socialisation and long-term relations in certain places in the everyday world, such as neighbourhoods. I see in the analysis of life-story interviews a possibility to reconstruct 'the multiple senses and meanings of place constructed by the co-inhabitants of any place' (McDowell, 1997, 2). Such a phenomenological view starts from the experience of place and not from the assumption of an empty and neutral space which is then divided into different places. Time and space 'remain, first and last, dimensions of place, and they are experienced and expressed *in place by the event of place*' (Casey, 1996, 38). It is obvious that neighbourhoods are a relevant point of reference in such a conception of place – how people co-constitute and experience them and, in this chapter on past neighbourhoods especially, what happens when and after they completely cease to exist. Emplacement can thus be defined as the piling up of perceptions and experiences of and in places that individuals co-constitute for shorter or longer periods. Emplacement might further be heuristically divided into 'emplacing oneself' – which places become relevant biographically in terms of experiences, the (changing) meanings ascribed to places and memories attached to them – and 'being emplaced' by being part of different figurations and by discourses about the places that individuals are part of.

However, as mentioned above, analysing biographical-narrative interviews allows not only the approximating or addressing of past experiences and actions, but also the perspective from which biographers speak in the present and how they frame and choose topics. In terms of place, the interviews may offer access not only to processes of emplacement in the past, but also to the way these are remembered and expressed in the present. *Memories of place –* or memories of emplacement – are never a replication of the experienced emplacement. Rather, they are determined not only by past experiences, but also by the current life situation of those who remember, and by the interview situation itself. Memories are also influenced by collective memory and discourses, as well as by the figurational positioning of the biographer, which contribute to the selection of and criteria for memory presentation in the present (Rosenthal, 2016b). For example, looking at the two cases that I have chosen to discuss here, it is clear that Huda and Jamal, as Palestinians, are in an outsider position in the figuration with Jewish Israelis in Jerusalem, in the sense proposed by Norbert Elias (Elias and Scotson, 2008/1965). They thus have less power to shape the memory of the former Palestinian neighbourhoods in a wider discursive arena. The Israeli definition of the former neighbourhoods as part of the 'Jewish Quarter' has become dominant even in the Palestinian community. This is visible in that the Israeli drawing of boundaries and naming practices have been widely accepted. However, the temporal and spatial process of marginalisation is also connected to intra-Palestinian power asymmetries, as I will show later.

Former Palestinian neighbourhoods and today's enlarged Jewish Quarter

The enlarged Jewish Quarter's history is an especially complex aspect of the Middle East conflict. Within its current, administratively defined boundaries live 2,900 people, according to statistics from 2014 (Jerusalem Institute for Israel Studies, 2016). The enlarged Jewish Quarter is usually identified as one of four spatially fixed ethno-religious quarters of Jerusalem (alongside the Christian, Armenian, and Muslim Quarters). The history and present of this ethno-religiously defined space are, however, much more complex than the seemingly clear spatial delineation suggests. Today, legislation makes it practically impossible for Palestinians to settle there. Historically, a homogeneous Jewish Quarter probably never existed. It wasn't exclusively Jews who lived there, and the Jewish inhabitants didn't form a homogeneous grouping. In adjacent areas of the Old City, there was a Jewish majority at times, but it was not called the 'Jewish Quarter'. The idea of four 'pure' ethno-religious quarters only emerged in Western and Zionist discourses during the late 19th or early 20th century.[3] Even today, there are over 500 Palestinians living in the enlarged Jewish Quarter.[4]

Before 1948, the area of today's enlarged Jewish Quarter consisted of different neighbourhoods (*ḥārāt*), the names and borders of which had changed throughout history. One of them was the *Ḥārat al-Yahūd* (referred to below as the Jewish Quarter). Others were the mixed, but majority-Muslim *Ḥārat aš-Šaraf* or the *Ḥārat al-Maġāriba* (Moroccan Quarter), which bordered the Western Wall. Many of the Moroccan Quarter's residents traced their origins to pilgrims from North Africa who had remained in Jerusalem. There were also other small neighbourhoods, the borders of which sometimes overlapped. The former Jewish Quarter was centred around the synagogues in the heart of today's enlarged Jewish Quarter. It also housed people belonging to other religious groupings. Dumper (2002, 80) holds that it is difficult 'to be precise about the exact location of the traditional Jewish Quarter because [...] the borders between all the quarters of the Old City fluctuated according to immigration and political circumstances of a given period'.

The Jewish Quarter was often described as a rather poor area. Because of the increased building activity outside the Old City walls from the middle of the 19th century, new city quarters were perceived as providing better living conditions than 'inside' the walls. Therefore, especially from the beginning of the 20th century onwards, the Jewish Quarter was 'associated with the old, poor, and weak' (Bar and Rubin, 2011, 776).

In the 30 years between the Jordanian conquest of the Old City in 1948, the occupation by Israel in 1967, and the end of the 1970s, the population was almost entirely exchanged, and the area was almost completely rebuilt. The Jewish Quarter fell to Jordan after a long siege by Arab armies in the war of 1948. Around 1,500 Jewish inhabitants fled or were expelled. Synagogues and many houses were damaged during the fighting, or destroyed later by the Jordanian army or local inhabitants. Palestinian refugees from the newly founded Israeli state, and especially poor internal migrants, mainly from the Hebron region, settled in the ruins of the empty houses (Bell et al., 2005, 18; Ricca, 2007, 52). Israel conquered the Old City in the war of 1967. One week later, the municipality and the Israeli government demolished the Moroccan Quarter adjacent to the Western Wall. Approximately 650 to 700 inhabitants were forced to leave. The area of the former Moroccan Quarter is today the large square in front of the Western Wall (Abowd, 2000, 12–13). After this, the authorities envisioned 'reconstruction' of the Jewish Quarter, a process that took more than 15 years.

In 1968, the Israeli Ministry of Finance expropriated 700 buildings, not only in the area previously identified as the Jewish Quarter, but also in surrounding neighbourhoods.[5] There are conflicting estimates regarding this enlargement, ranging from double to six times the area of the assumed former Jewish Quarter (IPCC, 2009, 11; Dumper, 1992, 37–38). As already indicated, the authorities did not base their conception on the idea of small, changing neighbourhoods, but assumed a grid of four objectively existing quarters in the Old City. For example, the judges of Israel's Supreme Court backed legislation to severely limit the inward migration of non-Jews to the newly defined area:

> Naturally, the reconstruction is aimed at restoring the former glory of the Jewish settlement in the Old City, so that the Jews will once again, as in the past, have their own unique quarter, alongside the Muslim, Christian and Armenian quarters. There is no wrongful discrimination in distinguishing these quarters, each quarter and its congregation.
>
> (Israel Supreme Court, 1978)

The expulsion of 5,500 to 6,000 Palestinians took place from 1969 until at least 1977. It remains unclear how many of them accepted the financial compensation offered. After the expulsion, most of the old buildings were demolished and new houses were erected. I was not able to find out why some Palestinian families were allowed to stay in their homes, or why their eviction was not enforced. An article in an Israeli architecture journal states, without any reference, that letting 25 families stay there was a gesture of good will on the part of then Prime Minister, Menachem Begin (Hattis Rolef, 2000). With regard to the 'new' inhabitants, there was an initial plan to settle a 'mix' of religious and non-religious Jews, but in the 2000s 95 per cent of the inhabitants were defined as religious (Glass and Khamaisi, 2005, 5).

The 'reconstruction' was based on a master plan designed in line with Western architectural principles. According to Simone Ricca (2007, 36), the architects re-interpreted the Jewish Quarter's past as glorious. They were not fond of the original 'oriental' and rather modest building style and wanted to incorporate 'Western rationality' into their buildings, as they thought appropriate for the Zionist idea. Although largely designed from scratch in the late 20th century, the present Jewish Quarter signifies authenticity for many Israelis and tourists. They identify it as a historical Jewish neighbourhood in the 'mosaic' of the Old City where every religious grouping has its own quarter.

Here, it is important to keep in mind that the enlarged Jewish Quarter differs considerably in its appearance from other areas of the Old City. For many, the atmosphere suggests a haven of calm and cleanliness in comparison to other, mainly Palestinian, Old City areas. Even for many Palestinians, the enlarged Jewish Quarter serves as an example of a more modern city. I heard this several times in my interviews, and Simone Ricca (2007, xiii) summarises it thus: '[T]o many Palestinians it is seen to be a successful model of urban reconstruction, to be eventually copied and imitated.' During my field research, it became clear to me that many Palestinians accept the enlarged Jewish Quarter, both its name and its reality. The names of the above-mentioned historical neighbourhoods have to a large extent disappeared from daily usage. Huda, for example, did not once use the Arabic name of the neighbourhood in which the house, where she still lives, was located. She uses the term 'Jewish Quarter'. For her, and for Jamal as well, the Israeli definition of the enlarged Jewish Quarter is dominant. There generally seems to be little interest in looking back at the past of the enlarged Jewish Quarter. Most neighbourhoods are in the process of being forgotten, and the Israeli definition of this space seems to prevail. I will address these aspects in the following sections.

Temporal marginalisation: lacking memories

Jamal is the father of my field assistant, Lina, who worked with me in the enlarged Jewish Quarter.[6] Lina was a 20-year-old student of psychology in West Jerusalem at that time, and well-informed about the city's history. She was much closer to her mother's side of the family than to her father's side. Her mother comes from a well-known and prosperous Jerusalemite family who still have their family home in the Old City. She was not aware that her father, Jamal, who comes from a poor family of internal migrants, grew up in the enlarged Jewish Quarter before leaving in the wake of the 'reconstruction' in 1970, when he was about 10 years old. This only became clear to her when I conducted a life-story interview with Jamal in her presence. Before our collaboration, Lina had also never visited the enlarged Jewish Quarter herself and did not know anything about it.

During the interview, however, Lina seemed largely uninterested in the fate of her father's family. This observation hints at one of the reasons why the memory of the Palestinian neighbourhoods is in the process of being forgotten within the Palestinian community. Jamal, like many who were evicted from this area, came from a poor family of internal migrants from Hebron. His family is in an outsider position in relation to Lina's Jerusalemite family. Before meeting Jamal, we had interviewed two other members of Lina's family. When I asked her if we could talk to more members of her family, she again tried to find people from her mother's Jerusalemite side. Only after unsuccessfully trying to persuade five other maternal relatives did she say that we could interview her father. Lina's mother, in our interview, told us that she now regretted their marriage, to which she had agreed despite opposition from her family. This shows a clear separation of the two sides of the family, with Lina's father's family seen as not fit for presentation to the outside world. This continued during the interview, during which Lina and her mother were present. They both commented on what Jamal talked about. Sometimes they pressured him to detail certain memories. But often, they whispered and laughed about Jamal in the background, especially when he was talking about his family of origin. I assume that they were attempting – though on a latent level – to distinguish their own background by ridiculing Jamal's.

Jamal's paternal grandparents had lived in Hebron's Old City. They died young, and subsequently their eldest son – Jamal's father – relocated to Jerusalem around 1947. He came with his wife and four brothers and found a one-room flat in a crowded inner courtyard in the *Ḥārat aš-Šaraf*, one of the neighbourhoods that now belong to the enlarged Jewish Quarter. Jamal's father opened a vegetable stall in the Old City. Both parents were poor and illiterate; they had barely been to school. Jamal, the family's second son, was born in 1961. During our interview, he took a lot of time to talk about the poverty he experienced within his family when he was young. He described the difficulties of a family of 11 living in one room, sleeping on the floor and being forced to share sanitary facilities with six other families. This description of his childhood prompted laughter from his wife and daughter, which perplexed me, as it seemed so inappropriate in that situation.

After the beginning of the Israeli occupation, the house Jamal's family lived in was expropriated by the Israeli Ministry of Finance. In 1970, Jamal's parents left the *Ḥārat aš-Šaraf*. They used the compensation money, which they accepted, to rent a small house in *Wadi Juz*, a neighbourhood outside the Old City walls. The following brief quote from the interview, which was interspersed with frequent breaks, is Jamal's summary of the eviction, and he did not say much more about it when I asked him for details later on.

> Then I went to school here, also we left [8 seconds silence] the Jewish Quarter in uh 68 uhm uh no in 70 we left the Jewish Quarter and we our parents moved to Wadi Juz [6 seconds silence], they left the Jewish Quarter because uh I remember that they they have to to clean the Jewish Quarter from the uh Muslim citizens there and some people took money, some left.

For Jamal, although he and his family were evicted, the forced departure from the *Ḥārat aš-Šaraf* did not constitute a theme in itself. Rather, it was embedded in his report of his school career. The quote shows very clearly that the eviction is not something he usually talks about. The long interruptions signify that he needed time to organise his thoughts and to find suitable words. Jamal does not have a practised way of talking about these events, nor does he have recourse to the language usage of a we-group of Palestinians from the former neighbourhoods in the Jewish Quarter, which would help him to explain how the family suffered from the Israeli policies. This is very different from the self- and we-presentations of Palestinian refugees of 1948, for whom there are discursively shared ways of talking about themselves (Worm, Hinrichsen, and Albaba, 2016). Jamal cannot resort to a shared and practised discourse to address his family's displacement. What is even more striking is that he uses the language of an Israeli discourse to talk about what he experienced with his family after 1967. Analysing the above quote closely, Jamal seems to have read publications written from an Israeli perspective in his search for information, as indicated by the terms he uses. While the term 'to clean' carries a racist connotation, the term 'leave' is a clear euphemism. His usage of the term 'Jewish Quarter' instead of the previous Arabic neighbourhood names points in the same direction. Overall, it looks as if there is no space for Jamal to address his past experiences when talking to others, even within his family. This can be attributed to the fact that he comes from a lower-class family of internal migrants seen as not having a long history of living in Jerusalem, something Jamal himself probably perceives, and as unfit for outside presentation. This puts him in an outsider position in the familial dialogue and prevents him from talking about the past. Jamal's family history occupies a marginal space in the family discourse. This suggests that he has been left alone with his memories of this period, which certainly also include experiences of violence and discrimination.

During our conversation, Jamal emphasised that he and his family 'were with the last citizens that left'. This was a recurring argument in the interviews I conducted with those who were evicted from the enlarged Jewish Quarter. It expresses *ṣumūd*, meaning steadfastness, the idea of holding out as resistance against the occupation (Farsoun and Landis, 1990, 28). However, the question of whether his family accepted the compensation offered was always in the air. After the interview had progressed for a while, Jamal admitted that his parents had taken the money. He added that, had it been today, or had they known more back then, they would not have moved from the neighbourhood so compliantly, but 'the Jews made it very good, very easy for us to leave at this moment'. By mentioning this detail, Jamal goes against the normal expectation of fighting the occupation with *ṣumūd*. Even though the Jerusalemite family of his wife is not known as a very nationalistic family, the fact that Jamal's parents (as many others) accepted the compensation must sound very shameful, given today's prevalent version of the patriotic discourse.

Jamal noted that he does not return to the enlarged Jewish Quarter to see what it is like today. Instead, while talking about the eviction of his family, he was constantly browsing through his mobile phone to find pictures of the area that he had downloaded from the Internet. When I asked him to tell me more about the specific neighbourhood he lived in (which he had barely mentioned, more often using the term 'Jewish Quarter'), he said: 'You see how *Ḥārat aš-Šaraf* looks now [...], now they have electricity, they have water, but before it was very dirty.' With this argument, he denigrates the atmosphere of poverty during his childhood, as seen from the present. From a reconstruction of other excerpts from the interview, it seems likely that he concealed other, more positive memories, or that he coloured them during the interview in his presentation of a poor childhood in a dirty neighbourhood. He compared his own past with the media images he found on the Internet, and interpreted the situation as being better now.

Jamal made no attempt at all to tell us that today this area does not belong to its rightful inhabitants, or to blame the Israelis for the evictions. Lina urged him: 'Speak, speak!' while Jamal continued to search for images on his mobile. Lina became increasingly annoyed, and finally burst out in Arabic: 'Johannes knows the area better than you!' This accusation likely showed not only how disappointed she was with what her father's family did in 1970. Maybe it was the depiction of his poor, depressing upbringing, or an unspoken accusation that Jamal is not a 'real Jerusalemite'. If his family had 'originally' been from Jerusalem, his daughter would not have asked him to prove his knowledge of the space, more or less as evidence that he has the right to speak about these issues. The limited ability of Jamal to talk about his life story and his family's eviction from the enlarged Jewish Quarter is connected to his outsider position within the family context, and more generally as a 'Hebronite' with less authority to speak about the city than 'real Jerusalemites'. The story of poor migrants who accepted compensation does not suit the narrative of a 'Jerusalemite family' with a century-long, uninterrupted history in the city. His story becomes irrelevant or even embarrassing to the family story.

The historical developments surrounding today's enlarged Jewish Quarter are not a common part of Palestinian collective memory, and – with the exception of the Moroccan Quarter – seem to be falling into collective oblivion. I know of no initiatives to preserve the history of these areas or to bring the voices of those who lived there to the fore. There are several possible reasons for this silence: In the *Ḥārat aš-Šaraf* and other, smaller neighbourhoods, the evictions were a slow process, spread over a period of at least eight years. The families who were forced to leave were generally poor, mainly refugees and 'Hebronites'. Some considered leaving as a form of relief, and saw the compensation Israel offered as quite generous in relation to their poor living conditions. The relatively large number of families who accepted the compensation is hard to understand for many people, as it is more frowned upon today than it was back then. All this plays a role in the fading of memories of the expulsion; of the Palestinians I talked with outside the Old City, not many knew of these historical developments. Almost all of them spoke of the 'Jewish Quarter' when referring to the area defined by the Israeli administration. Names and the historical developments surrounding the former Palestinian neighbourhoods in the enlarged Jewish Quarter have not entered Palestinian collective memory.

Spatial marginalisation: lacking neighbourhood life

For the emplacement of those Palestinians who remained in the enlarged Jewish Quarter it is important to consider – in phenomenological terminology – whether they continue to be presented with a noematic system of Palestinian neighbourhoods. In other words, whether the neighbourhoods that presented themselves to their inhabitants as given in the past still exist as a category, even if the make-up of individuals and things has fundamentally changed. This raises the question of whether the 'remnants' – the remaining Palestinians and their houses – are sufficient to identify the structure of the 'whole' neighbourhood (cf. Husserl, 1982; Becker, 2017a). In the case of the Jewish Quarter, my analysis suggests that these neighbourhoods are no longer traceable. They have more or less ceased to exist and can be seen as being in the spatial past. I will detail this by introducing my analysis of Huda's biography.

In 2012, I conducted a 6-hour interview with Huda, a young Palestinian woman living with her parents and siblings in the enlarged Jewish Quarter, meeting her three times.[7] Huda was born in 1987, one year after her brother and one year before her younger sister; two younger brothers followed later. The family house is in close proximity to Israeli neighbours and isolated from other buildings occupied by Palestinians. The area is heavily frequented by tourists and visitors, and police are stationed nearby.[8] Huda's family has been living in this building for at least three generations; it is privately owned by her father's family. Its prominent position in the enlarged Jewish Quarter has two interconnected effects for the family. On the one hand, the family has to fight attempts by Jewish religious nationalists and neighbours to bring the house into their possession. On the other hand, there is the Palestinian community's expectation that the family must remain there against all odds. In the light of the growing number of Jewish settlements, Jordan, and later the Palestinian Authority, threatened those who sold their homes to Israelis with the death penalty.

Huda's father is the last of his family to live in the building, his brothers having moved out of the Old City when Huda was a child. This means that Huda's nuclear family has an isolated position. As 'those who remained', they are ascribed the societal task of retaining the building, although her father's siblings had attempted to sell it. This would have meant immediate wealth, but also the family's complete exclusion in Jerusalem. These attempts gave rise to serious conflicts within the family. Jewish activists and neighbours also pressured the family to leave by means of psychological and physical violence. They attacked Huda and her brother several times in the streets, took illegal photos of the inside of the house and the family members, assembled loudly on the roof of their building, and offered large amounts in payment. These experiences have been part of Huda's everyday experience since her childhood.

However, in Huda's biography, the theme of the family's exclusion in the enlarged Jewish Quarter, her perception of strangeness or isolation, is strongly connected to another aspect: Huda's restriction by her nuclear family. Although related to the family's exclusion, Huda blames her family for confining her to the family house – this forms a recurring theme in the interview. It is related to Huda's attempt in the present to distance herself from her family, and shows a precarious balance that is also typical of Huda's biography: Huda has increasingly found herself caught in a balancing act between belonging to the we-group of the family in the house, belonging to the we-group of Palestinians, and having to deal with her family's difficult position in the enlarged Jewish Quarter. While Jamal's difficulty in talking about his childhood is determined by his outsider position as a poor internal migrant who married into a Jerusalemite family, Huda's current outsider position within her family is constituted by her attempt to overcome gender-based control and social disempowerment.

During her childhood, Huda and her siblings played in a large square in the Jewish Quarter, which is hard to imagine in today's political situation. But

Huda tells me that it was suddenly 'forbidden to step out of the doorstep, we have to always sit in the same house and play there'. While it would be obvious to blame the political situation for restricting neighbourhood life, Huda places the responsibility on her mother. Even more, Huda perceived this restriction as a continuous state. In the interview, she skips several years within a few seconds and continues by talking about her time at school, when she experienced the same strict routine every day:

> We went [to school] on time and we returned on time, there was no time to play and no place where we could have gone, we returned and after a while the night came [...], my mother was even afraid of youth clubs, afraid of every place, place, no place.

Attempts to play in public spaces near their home often ended in being attacked by the neighbours' children, and subsequently in being confined to the family house by her mother. Playing in the car-free alleys of their neighbourhoods is among the positive constitutive childhood experiences mentioned by most other interviewees. By contrast, Huda says: 'I hated the streets and the Old City maybe because we were outsiders in a way, in other words I was not really involved in the life of the city.' The members of her family were outsiders in the quarter in relation to their Jewish neighbours, but also outsiders within the Palestinian Old City community. Most narrations about her childhood concern the restrictions she experienced in both her nuclear family and the enlarged Jewish Quarter. Many other narrations show that Huda's perception of her mother as the 'custodian' of the family house largely overlaid negative experiences with Jewish neighbours or activists. Although her narrations of their often aggressive behaviour show her fear, they reveal more agency than stories about her nuclear family:

> We forgot the door open so we saw how many people entered [...], I took the broom and we were shouting at the Jews to leave the house [...], after they left the house I felt victory, but at that time we were very scared [...], I was telling this story to everyone so I felt again that I won this time.

Even though the pressure in the quarter created familial problems, at the same time it also formed a bond between the family members. When Huda talks about hostility towards her family, she uses the personal pronoun 'we': 'If they want to take the house they can just kill all of us and bury us in front of the gate and they can take it, but we're not going to leave this house.'

Huda started going to a Christian girls' school run by European nuns in the Old City in 1993. Huda does not talk about her school days much, but continues to concentrate on her domestic restriction. The beginning of her teenage years coincided with the Second Intifada. This started in September 2000, and came along with clashes in the Old City when Huda was 13 years old. At this time, she started to develop a political awareness, criticised her parents' political passivity, and accused them of keeping their heads down. Huda started to link this position to the restriction exercised by her parents. She recognised that she was spatially separated from other Palestinians in the Old City, that she attended a Christian school managed by foreign nuns, and that this prevented her from developing a Palestinian 'political and national identity'. She increasingly saw herself as being in the wrong places. She says: 'I was not allowed to express my feelings at home, but I was also not allowed to express my feelings at school.'

A decisive date for Huda was when she changed to a public Palestinian school in 2003/2004, at the age of 16. She described this as her attempt to 'be in an environment where I belong'. She wanted to be integrated into a 'normal' Muslim-Palestinian majority environment, and to deal actively with her spatial position. She also started wearing the hijab, not for religious reasons, but as a symbolic affirmation of this move. For Huda, these steps were related to acquiring a feeling of agency and self-confidence: 'From that time, I felt that I don't really need my family and I became less scared from facing things.' However, when she began studying primary education after passing her final high school examination (*tawjihi*), daily control by her family did not decrease. Her elder brother played an increasingly important role in this, and control was also connected to physical violence (cf. Joseph, 1999).

During her time at university, she studied the city's history extensively, took part in field research in the Old City, and volunteered in a local children's centre. After attending a theatre workshop, she performed her own programme at local girls' schools. Huda presented 'typical' Jerusalemite characters and introduced different Old City neighbourhoods. She tried to incorporate this knowledge as a creative voice of the Palestinians in Jerusalem's Old City. Huda says that she had previously only focused on the problems of poverty and backwardness in the Palestinian Old City neighbourhoods. She compared this with the positive image of the Jewish Quarter, 'which is clean, and people somehow treat each other in a good manner'. Having learnt about the importance of the Old City, however, Huda recognised her own special position there. She filled it with meaning for her own biography, which she then linked to her personal fate of being placed in the city. Huda emplaced herself in Jerusalem because of its symbolic significance. Although her everyday restriction in the quarter and family home did not change, her perception of isolation in the enlarged Jewish Quarter became modified. She said she had become a 'friend of Jerusalem'. Maybe it was the fragmentation and precarious environment of Palestinians in the Jewish Quarter and the connected lack of communalising structures and public space that led Huda to re-connect with Jerusalem as a whole as an alternative emplacement.

After completing her degree, Huda was employed by a girls' school in the Old City. She was still severely controlled by her family. Her increasing attempts to be independent and to move within public spaces angered her family and triggered violent repercussions. After an especially harsh physical attack by her brother, Huda escaped from the house and ran towards the Israeli policemen stationed nearby:

I ran into the quarter because he hit my eye and my head very very much and police, suddenly everything was police [5 seconds silence], this is the first time, first time, first time that I feel that that my enemy has saved me from my family.

It was especially difficult for Huda to talk about the appearance of the police in the house. Violence was an everyday experience for her, but the intermingling of family home and Jewish quarter made this situation different. Events in the quarter often triggered familial problems at home, but in this instance, the normal chain of events and the separation of quarter and home broke down.

In this period – around the time of the interviews – Huda's symbolic emplacement in Jerusalem became even more important as a way of explaining and understanding her life situation. She repeatedly uses the metaphor 'Jerusalem' to individualise herself and to put herself on one level with the city, or to present herself as 'united' with the city:

When they asked me who I am I say I am Jerusalem […] with all details, with its alleys, with their twists, with its simplicity, its history and venerability […], I look like Jerusalem with its sadness, brokenness and its defeats.

Using this metaphor, she was able to fit her biographically difficult position into the fate of the city; she herself became part of its national and religious importance. Her growing awareness that she was living as a woman in a special 'place' inside the already charged Old City contributed to her interpreting herself as a representative of the city. The symbolic emplacement helped her to make sense of her own biography, to work through her biographical problems, and, to a certain extent, to distance herself from her family.

Conclusion

I have argued that the process of destruction of the neighbourhoods in the area of today's enlarged Jewish Quarter in Jerusalem did not end with the eviction of the residents, but that a process of marginalisation has continued on two levels: spatial and temporal. These different, but interdependent, processes are exemplified on the one hand by the fading, isolated memory of Jamal whose family was evicted, and on the other by the lack of neighbourly space for Huda, whose family remained. In the enlarged Jewish Quarter, for both Jamal and Huda, the former Palestinian neighbourhoods have disappeared and their boundaries have vanished. Thus, the neighbourhoods have lost qualities such as being a symbol of security and social control, or being a (positive or negative) reference point, leading in Huda's case to her restriction to the family house. Failing to offer mutual assurance, they ceased to be places for emplacement or that offer a we-image for the Palestinians who remained there. The neighbourhoods are part of the past, they do not play a role in everyday life, and memories of them are fading. While marginalisation is mainly triggered by Israeli politics of space, it is also connected to intra-Palestinian power dynamics between the established and outsiders (Elias and Scotson, 2008/1965), for instance, between internal migrants and long-established families, or between men and women.

From the perspective of sociological biographical research and based on the notion of the mutual constitution of places and people, I have reconstructed Huda's *life history in order to analyse her biographical experiences of place as emplacement* – the piling up of perceptions and experiences of and in places to which individuals 'belong' and which they co-constitute for shorter or longer periods. In my analysis of Jamal's *life story, I have reconstructed his memories of place in the present of the interview*, and how these are interrelated with collective memory, discourses, and his figurational outsider positioning. Memories of places are never constant objects. By analysing the current positioning from which they speak, both can be seen as belonging to an outsider grouping of Palestinians in Jerusalem, with little chance to influence the hegemonic discourse on the enlarged Jewish Quarter. Both can also be seen as being in an outsider position within their families – Jamal as the 'Hebronite' in a Jerusalemite family, and Huda as an ambitious young woman in a patriarchal, conservative family setting. This influences how they approach their memories, but in different ways.

Jamal has neither the discursive nor the familial means to speak about his childhood in the enlarged Jewish Quarter and his family's eviction when he was 10 years old. Within the established Jerusalemite family he married into, his origin in a poor family of internal migrants is usually de-thematised in the family dialogue, as shown by the organisation and course of the interview. Furthermore, he does not possess pictures or any other mnemonic materials to support his memory of the neighbourhood he lived in as a child. He hardly ever talks about it, and the old names and borders have vanished, so that he resorts to Israeli images showing the current layout of the quarter, which does not match his memories. His memories have become merged with the Israeli discourse of dirty neighbourhoods that were cleaned up. His ability to recall, and especially to communicate, his memories gets weaker over time. Even within the range of communicative memory (Assmann and Czaplicka, 1995), the former neighbourhoods seem to be in the process of being forgotten. There is little shared knowledge and, for Jamal, no counterpart to talk with about them.

In the case of Huda, the neighbourhood is restricted by her problematic emplacement in an isolated family house. The often-hostile environment in the administratively defined enlarged Jewish Quarter has led the importance of this emplacement to grow. The dearth of neighbourhood life and of available public spaces became problematic, especially in the context of primary socialisation and familial control. The Palestinian social (and legal) pressure to retain the family house, and the attacks by the neighbours, increased the family's perception of their separation. This contributed to the genesis of familial conflicts and reinforced Huda's confined emplacement in the family house. Huda tried to work through this confinement by seeking a Palestinian majority environment, for instance by changing schools, as well as through her studies of Jerusalem's history and present. This culminated in her symbolic emplacement in the Old City by her framing her situation as a special closeness to the city and its history.

Fragmentation and 'islandisation' in family houses were reoccurring themes during my field research in today's enlarged Jewish Quarter. The lack of neighbourhood or other communalising structures, and the lack of neighbourhood life for Palestinians, mean that public space is problematic and social control is less strong than in other areas of the Old City. Therefore, social life differs. The lack of neighbourhood structures in connection with the precarious surroundings creates a need for alternative emplacement, as exemplified by Huda's symbolic re-connection with the city. Restriction to the family house and alternative emplacement can also be read as a shift towards (forced) individualisation, especially in comparison to other Palestinian neighbourhoods in the Old City. There, as my fieldwork revealed, emplacement in the neighbourhood is still dominant as a (positive or negative) reference point. Neighbourhood belonging continues to offer mutual identification and assurance. It can be used as a self-ascription of belonging and as an emplacing term. For example, the small community-oriented neighbourhood within which I conducted another part of my fieldwork still presents itself as a possible we-image to its inhabitants. To the outside, many residents introduced the neighbourhood as a discursively shared ideal of community. Without delving into the discussion as to the extent to which neighbourhood residents are free to follow or reject the lifestyle that comes with living there, I can say that this neighbourhood symbolises both a source of security and considerable social control, although it does not determine emplacement *per se*, and allows, for instance, for retreat into a private environment, or part-time detachment from the shared lifestyle (Becker, 2016). By contrast, the former neighbourhoods in the enlarged Jewish Quarter have been placed firmly in the past by two different levels of a process of marginalisation – temporal and spatial. This prevents them from being lived or remembered in the present. Loss of memories, a lack of memory resources, and familial disinterest limit people's ability to talk about the pre-eviction neighbourhoods in the present. They do not play any constitutive role in everyday life; the spatial environment is determined by inimical neighbours and the burden of safeguarding the family's property.

Eviction or physical destruction proved insufficient to place neighbourhoods in the past. It was the subsequent marginalising processes that I have reconstructed here that resulted in the discursive hegemony of the Israeli spatial history and the successful isolation of those who remain. Taken together, it led to the inability to identify the former neighbourhoods on the basis of their remnants, guaranteeing that they remain firmly in the past.

Notes

1 The other two spaces being a small neighbourhood in a mainly Muslim area, and international monks residing in the Old City. My research was conducted in the framework of the project 'Belonging to the Outsider and Established Groupings: Palestinians and Israelis in Various Figurations', funded by the German Research Foundation (DFG) from 2010 to 2015 (see Rosenthal, 2016a). During my research trips, which totalled eight months altogether, I conducted 35 biographical-narrative interviews, as well as short- and long-term participant observation.
2 Which could also be read as a shift in the 'we–I balance' (Elias, 2010/1987), from a more communalising 'we' in the direction of an individualising 'I'.
3 For a more detailed introduction, see for example, Arnon (1992), Tamari (2000), Wallach (2011) and the summary in Becker (2017a, ch. 6).
4 The most recent figures are from 1998: 480 Muslims and 17 Christians. See Dumper (2002, 29), Ricca (2007, 213, note 32).
5 In 1948, before the war, 105 of these buildings had Jewish owners, 130 were in the private possession of Palestinians, 354 were registered as private and 111 as public Muslim *awqāf* (Ricca, 2007, 50–52).
6 Names and other details have been changed to ensure anonymity. The histories of Lina's mother's and father's families are described in detail in Becker (2017b).
7 Huda was introduced to me by a colleague who was working on the same research project. He also joined me for the interviews. The quotes in this chapter are either a transcript of the English spoken during the interview or are based on my own translation of the Arabic interview transcript.
8 For reasons of anonymisation, I cannot describe the exact position of the building.

References

Abowd, Thomas. 2000. 'The Moroccan Quarter: A history of the present'. *Jerusalem Quarterly* 7: 6–16.
Apitzsch, Ursula and Lena Inowlocki. 2000. 'Biographical analysis: A "German" school?'. In Chamberlayne, Prue, Joanne Bornat and Tom Wengraf (eds.): *The turn to biographical methods in social sciences: Comparative issues and examples*, 53–70. London: Routledge.
Arnon, Adar. 1992. 'The quarters of Jerusalem in the Ottoman period'. *Middle Eastern Studies* 28(1): 1–65.
Assmann, Jan and John Czaplicka. 1995. 'Collective memory and cultural identity'. *New German Critique* 65: 125–133.
Bar, Doron and Rehav Rubin. 2011. 'The Jewish Quarter after 1967: A case study on the creation of an ideological-cultural landscape in Jerusalem's Old City'. *Journal of Urban History* 37(5): 775–792.
Becker, Johannes. 2016. 'Commitment to the Old City and ambivalent emplacement'. In: Rosenthal, Gabriele (ed.): *Established and outsiders at the same time: Self-images and we-images of Palestinians in the West Bank and in Israel*, 125–146. Göttingen: Göttingen University Press.
Becker, Johannes. 2017a. *Verortungen in der Jerusalemer Altstadt: Lebensgeschichten und Alltag in einem engen urbanen Raum*. Bielefeld: Transcript.
Becker, Johannes. 2017b. 'A mixed family of long-time residents and internal migrants in Palestinian Jerusalem: The established-outsider figuration of 'Hebronites' and 'Jerusalemites' in the context of the Israeli occupation'. In: Bogner, Artur and Gabriele Rosenthal (eds.): *Biographies in the global South: Embedded in figurations and discourses*, 209–235. Frankfurt: Campus.

Bell, Michael, et al. 2005. *The Jerusalem Old City initiative discussion document: New directions for deliberation and dialogue.* Toronto: Munk Centre for International Studies, University of Toronto.

Casey, Edward S. 1996. 'How to get from space to place in a fairly short stretch of time: Phenomenological prolegomena'. In: Feld, Steven and Keith Basso (eds.): *Senses of place*, 13–52. Santa Fe: School of American Research Press.

Dumper, Michael. 1992. 'Israeli settlement in the Old City of Jerusalem'. *Journal of Palestine Studies* 21(4): 32–53.

Dumper, Michael. 2002. *The politics of sacred space: The Old City of Jerusalem in the Middle East conflict.* Boulder: Lynne Rienner.

Elias, Norbert. 2010/1987. *The society of individuals.* (The collected works of Norbert Elias, vol. 10), Robert van Krieken (ed.). Dublin: University College Dublin Press.

Elias, Norbert and John L. Scotson. 2008/1965. *The established and the outsiders.* (The collected works of Norbert Elias, vol. 4), Cas Wouters (ed.). Dublin: University College Dublin Press.

Farsoun, Samih K. and Jean M. Landis. 1990. 'The sociology of an uprising: The roots of the Intifada'. In: Nassar, Jamal and Roger Heacock (eds.): *Intifada: Palestine at the crossroads*, 15–35. New York: Praeger.

Glass, Joseph B. and Rassem Khamaisi. 2005. *Report on the socio-economic conditions in the Old City of Jerusalem.* Toronto: Munk Center for International Studies, University of Toronto.

Hattis Rolef, Susan. 2000. 'The Jewish Quarter in Jerusalem'. *Architecture of Israel Quarterly* 39. Accessed: 10.02.2019. www.aiq.co.il/pages/articles/39/jerusalem.html.

Husserl, Edmund. 1982. [German 1913]. *Ideas pertaining to a pure phenomenology and to a phenomenological philosophy: First book: General introduction to a pure phenomenology.* London and The Hague: Nijhoff.

International Peace and Cooperation Center (IPCC). 2009. *Jerusalem: The Old City: The urban fabric and geopolitical implications.* Jerusalem: International Peace and Cooperation Center (IPPC). Accessed: 22.02.2019. www.ipcc-jerusalem.org/attach ment/15/IPCC_Jerusalem_the_Old_City_Urban_Fabric_and_Geopolitical_Implica tions.pdf

Israel Supreme Court. 1978. *Burkan vs the Minister of Finance et al*, HCJ 114/78. Accessed: 13.02.2019. www.hamoked.org/Document.aspx?dID=Documents1204

Jerusalem Institute for Israel Studies. 2016. *Statistical yearbook, 2016 edition, table III/13, population of Jerusalem, by age, quarter, sub-quarter and statistical area, 2014.* Accessed: 21.12. 2019. https://jerusaleminstitute.org.il/wp-content/uploads/2019/06/shna ton_C1316.pdf

Joseph, Suad. 1999. 'Brother-sister relationships: Connectivity, love, and power in the reproduction of patriarchy in Lebanon'. In: Joseph, Suad (ed.): *Intimate selving in Arab families: Gender, self, and identity*, 113–140. Syracuse: Syracuse University Press.

Massey, Doreen. 1994. 'A global sense of place'. In: Massey, Doreen (ed.): *Space, place and gender*, 146–156. Cambridge: Malden.

Massey, Doreen. 1995. 'Places and their pasts'. *History Workshop Journal* 39(1): 182–192.

McDowell, Linda. 1997. *Undoing place?: A geographical reader.* London: Arnold.

Ricca, Simone. 2007. *Reinventing Jerusalem: Israel's reconstruction of the Jewish Quarter after 1967.* London: I.B. Tauris.

Rosenthal, Gabriele. 2004. 'Biographical research'. In: Seale, Clive et al. (eds.): *Qualitative research practice*, 48–64. London: Sage.

Rosenthal, Gabriele (ed.). 2016a. *Established and outsiders at the same time: Self-images and we-images of Palestinians in the West Bank and in Israel*. Göttingen: Göttingen University Press.

Rosenthal, Gabriele. 2016b. 'The social construction of individual and collective memory'. In: Sebald, Gerd and Jatin Wagle (eds.): *Theorizing social memories: Concepts, temporality, functions*, 32–55. London: Routledge.

Schütze, Fritz. 1983. 'Biographieforschung und narratives Interview'. *Neue Praxis* 13(3): 283–293.

Schütze, Fritz. 2014. 'Autobiographical accounts of war experiences: An outline for the analysis of topically focused autobiographical texts: Using the example of the "Robert Rasmus" account in Studs Terkel's book "The Good War"'. *Qualitative Sociology Review* 10(1): 224–283.

Tamari, Salim. 2000. 'Jerusalem's Ottoman modernity: The times and lives of Wasif Jawhariyyeh'. *Jerusalem Quarterly* 9: 5–27.

Wallach, Yair. 2011. *Shared space in pre-1948 Jerusalem: Integration, segregation and urban space through the eyes of Justice Gad Frumkin*. Divided Cities and Contested States Working Papers, 21.

Worm, Arne, Hendrik Hinrichsen, and Ahmed Albaba. 2016. 'The homogenizing we-discourse and the social positioning of the refugee camps'. In: Rosenthal, Gabriele (ed.): *Established and outsiders at the same time: Self-images and we-images of Palestinians in the West Bank and in Israel*, 45–66. Göttingen: Göttingen University Press.

6 Where is Alexandria?

Myths of the city and the anti-city after cosmopolitanism[1]

Samuli Schielke

The Chinese Housing

In March 2015, on one of my many journeys between Berlin and Alexandria, I landed at Borg El Arab airport west of Alexandria late at night. The airport is 50 kilometres away from the city centre, but close to many thriving industrial areas, holiday villages, and upmarket suburbs that have been built west of the city and on the North Coast in the past two decades. They are part of a general scheme by the Egyptian government to create new cities far from the old urban centres.

At the airport, I was picked up by my friend Mustafa, whom I have known since the days when he was still living in his native village in the northern Nile Delta. He has been living half an hour from the airport in the district of Agami at the western edge of Alexandria since 2009. Agami is known among the Egyptian bourgeoisie as a pleasant, traditional, and exclusive beach resort (Abdel Gabbar 2013). Mustafa, however, lives 3 kilometres away from the coast in an informally built area on a small hill right behind the Chinese Housing (al-Masakin al-Siniya), an ill-reputed area of large public housing blocks. The area was built in the 1980s as company housing for public sector companies by an Egyptian-Chinese joint venture.

For decades, the Chinese Housing had been an area where poor and marginalised people lived, people who lacked the means to build a house of their own in an informal settlement. It had experienced periods of gang wars that forced the inhabitants to stay indoors for up to three days. Since then, though, the neighbourhood had become calmer and the population more mixed. Mustafa and I moved through the area with no sense of risk even late at night.

Mustafa later told me that he initially didn't like the neighbourhood but grew to appreciate it for the opportunities it offered: 'For me, the Chinese Housing was like America.' He had just opened a shop in the area and business was reasonable. He said that unlike the commercial district of al-Manshiya in the historical downtown of Alexandria, the Chinese Housing and the surrounding informal neighbourhoods were not yet solid, not yet occupied. For Mustafa, it carried some of the mythological aura of the American dream.

Two years earlier, an Egyptian employee at a foreign research institute in Alexandria had been shocked to hear that I frequented the Chinese Housing. She said that she was surprised that I was still alive. For her it was a no-go

Figure 6.1 The Chinese Housing.
Source: Photo by Samuli Schielke (2016).

area, definitely not a part of her city. If anything, it was an *anti-city* neighbourhood that marked the boundaries of and threatened a bourgeois Alexandria, a cosmopolitan seaside city.

The next evening, I continued my journey on a minibus to the opposite end of the city, the neighbourhood of al-Mandara in the east. Al-Mandara is where I lived during my stays in Alexandria as a guest of the novelist Mukhtar Shehata, with whom I worked on an ethnography of literary writing, until 2017. To evade the congested roads along the city's seaside, the minibus takes a detour inland via the International Road that crosses Lake Maryutiya on a land fill bridge where the nauseating smell of pollution occasionally compels passengers to hold their noses. The road passes poor informal areas in inland Agami, the upmarket suburb of King Marriot, vast chemical and cement factory complexes, and the upmarket City Centre Alexandria shopping mall (not near the historical centre of the city). Finally, the minibus enters the city again along 45 Street, in what is known as the East of the City (Sharq al-madina). Approaching the end of the line, the minibus turns into smaller streets, passes the Faculty of Islamic Studies of the al-Azhar University, and finally enters busy Mallaha Street which is surrounded by shops, market stands, and congested by private cars, taxis, minibuses, and toktoks.

Eastern Alexandria is symbolically divided class-wise by the Abu Qir suburban train line, the seaside area being generally more well-off, and the inland area being mostly poorer. I got out at a spot where this mythological division is a tangible reality: at a minibus station next to the railway line. On the other, wealthier side of the railway are the Muntazah Gardens (formerly the royal summer residence, now a public park), the Fathallah shopping mall, the Sheraton, and the beach. On this side, the informal area of inland al-Mandara begins, where construction has been ongoing since the 1990s, with 15-storey houses replacing older five-storey ones.

In Mukhtar's words, this is 'the ugly face of Alexandria'. And it would be difficult indeed to find the Chinese Housing, the International Road, or inland al-Mandara beautiful in any conventional sense. It is not simply the poor face of the city, however. The suburban crescent that surrounds the old coastal core of Alexandria is made up of poor, middle-income, and upmarket neighbourhoods alike. They provide homes and work for millions, and yet none of these would count as the real Alexandria in the media, literary, and scholarly accounts of the city – and many of the inhabitants of the suburban crescent would agree. When I asked Mustafa what the real Alexandria is for him, he did not name the Chinese Housing where he lives, but rather the popular quarter of Bahary in the ancient centre of the city, his favourite site for outings with his family. Where, then, is Alexandria?

Alexandria has a reputation for being cosmopolitan – or having once been so, in a past *belle époque* when Europeans dominated the bourgeois districts of the city. That era is gone, but the reputation and romance of Alexandria live on. Western readers are likely to know Alexandria from the works of Greek, British, and other European writers who lived in the city, or more specifically, in the European-dominated central districts of the city that still carry the material memory of that era. Readers of those works will remain largely ignorant, however, of the vast majority of the city's inhabitants, Arabic-speaking Egyptians, and of the neighbourhoods they inhabited. Arabic literature on Alexandria is only gradually finding its way into the canon of Alexandrian cosmopolitanism (see Hazem 2006; Kararah 2006; Starr 2009; Halim 2013). Historians and literary scholars have provided textured accounts that question the Euro-centrism of the myth of cosmopolitan Alexandria (Fahmy 2006; Zubaida 2011: 131–155; Halim 2013; Chiti 2016; Hanley 2017). And yet an ambiguous nostalgia for a bygone cosmopolitan era is also shared by many Arabic-speaking inhabitants of Alexandria today, in a time when the city has left that era behind, and something rather different is emerging in a sweeping movement of urban erasure and expansion.

Which old and new myths of the city, I ask, are being crafted, questioned, or revised in such a moment, and what might they tell us about the wider imagined and material locations of the city? How do specific neighbourhoods figure in those myths? What political and moral claims about the city are involved in them?

Based on ethnographic fieldwork with contemporary Alexandrian writers and literary circles, I trace a selection of contemporary afterlives of the myth of Alexandria as something different and better than the actually existing city. The result is part ethnography of a specific literary milieu and part urban ethnography, where writers from that milieu provide the main theoretical inspiration. For an anthropologist, literature is not an obvious choice for understanding how a city is imagined and understood by its inhabitants. However, I encountered the question about the city through my ethnographic fieldwork with writers in Alexandria. The city is a major theme for many of them, and I needed to pay attention to that. Writers also participate in the wider popular cultural imagination as crafters of popular myths. Sometimes their works gain wider circulation when they are taken over by script and song writers, and by national and international cultural institutions. Some of these contemporary literary accounts of the city are fairly well-known nationally and internationally, such as Ibrahim Abdel Meguid's (1999, 2005, 2013) *Alexandria Trilogy*, and Alaa Khaled's (2012) literary work as well as his ongoing editorship of the journal *Amkenah* since 1990. There are many others who are less prominent but not less interesting. In this article, I engage with a handful of writers of the latter kind. All of them are more or less involved in the small, internationally connected and funded avant-garde scene in the city, a scene that is open to the world but limited in its societal reach (for other scenes, see Schielke and Shehata 2016).

When such rather cosmopolitan circles become the breeding ground for an intellectual critique of nostalgia for the cosmopolitan, something important is going on. Equally important is the historical moment in which such critique has emerged: the aftermath of Egypt's January 25 Revolution, which despite its political defeat has transformed both the way many young writers and intellectuals see the world and city they live in, as well as the material shape of the city itself. Since 2011, it has been subjected to rapid erasure and reconstruction in the course of a construction boom.

I work with concepts and ideas that I have encountered in my fieldwork, and consider them as theories that may or may not provide valid answers to the inquiry. I nevertheless call those theories myths because that is the form in which they circulate: as narrative, dramatic structures that may be told in different words (Lévi-Strauss 1955), and that naturalise moral and political claims and relations of power (Barthes 1970).

Alaa Khaled evokes the dreaminess of an Alexandria haunted by its myths – but also points out that those myths may change along with the city:

> The dream that was once planted into the consciousness of the city, will haunt it like a restless ghost, until it either takes material shape and returns to life, or this dream comes to an end and dies, or a new dream is invented.
>
> (Khaled 2012: 20)

Myths thus understood are compelling narrative structures that are to be judged by their power to inspire one to think and act along the lines they suggest. They have historical, political, and social lives worthy of attention (Chiti 2016), which means that they are never separate from struggles and relations of power, as pointed out by Roland Barthes (1970: 72): 'There are thus very likeable dreams which are however not innocent.' Following up with Khaled's, Chiti's, and Barthes' insights about the historicity and complicity of myths, I add that, when considered as social theories, some myths may also be better suited to providing guidance in a given reality than others.

Despite being highly mythologised, contemporary Alexandria is a rather ordinary city (El Chazli 2018), and its recent development is not remarkably different from so many other cities in the Global South that, in a short time, have transformed into vast conglomerates that have little in common with the cities they once may have been (see, e.g., AlSayyad and Roy 2005; Robinson 2006; Simone and Pieterse 2017). The nostalgia for a past colonial-cosmopolitan era is also a common feature of cities in the Global South (see, e.g., Bissell 2005; Newcomb 2017).

As Setrag Manoukian (2012) shows in his work on the Iranian city of Shiraz, talking about a city always involves highlighting some of its districts and neighbourhoods, and silencing others. Myths that purport to reveal a city's true location and values need to account for those locations and values that don't fit into the story: neighbourhoods and ways of living that are marked as not legitimately part of the mythologised city. I call them the *anti-city*. When I speak about 'the city' in the following sections, it is thus within the tension between the urban conglomerate that is too large for a textured account, my specific knowledge of some parts of it, and various myths that tell us what and where that conglomerate really is and ought to be – as well as what it is not and ought not to be.

Cyprus

Whenever I ask people where the 'real Alexandria' is located for them, I typically get seaside replies. They differ in terms of class (between the old popular neighbourhood of Bahary at the tip of the peninsula, the historical downtown area of al-Manshiya and Ramleh Station, and the old middle-class neighbourhoods east of downtown) and in terms of interests (between literates and summer guests), but there is wide agreement about the shore of the Mediterranean being Alexandria's proper location. And yet, over the past hundred years, the city's linkages with the inland have proved more enduring.

Founded by Alexander the Great in 331 B.C., Alexandria today bears few visible traces of its long history. Contemporary Alexandria is a child of the trade expansion and industrial revolution of the late 19th and early 20th centuries. Its rapidly growing population was mainly made up of migrants from upper Egypt, Nubia, and the nearby countryside, along with large numbers of European and Ottoman subjects who had moved there. The Alexandria of the colonial era was also a city of enormous inequalities and conflicts, and it could only last as long as the privileged position of foreign nationals lasted. Following the 1956 Suez Crisis, most Egyptian Jews as well as French and

British citizens were pushed to leave. The already dwindling Greek and Italian communities were allowed to stay, but the majority of their members gradually emigrated following the socialist nationalisation policies of the 1960s (Kazamias 2009). Alexandria became a city dominated by Arabic-speaking Egyptians of Muslim or Christian faith. They or their ancestors were once newcomers to the city, too, having arrived as rural–urban migrants in the city and having gone on to reproduce the plurality of Egypt within it.

And people keep arriving: rural–urban migrants and commuters, refugees from Syria and Libya, students from across the world studying at the Islamic al-Azhar University, Egyptian, Arab, and fewer Asian and Western tourists. But that mixture does not strike Western visitors and journalists as cosmopolitan. In Alexandria, 'cosmopolitanism' is usually equated with urban coexistence across religious and ethnic lines, but not all coexistence counts. Alexandria's cosmopolitanism is mainly equated with Europeans and European-dominated quarters.[2] This Eurocentrism is notably reproduced in the way 'cosmopolitanism' is used in Arabic as a French/English loan word: *kuzmubulitaniyya*, although the concept would be easily translatable into Arabic (Raouf 2016).

No wonder, then, that anything that happened after the 1950s hardly counts in standard accounts of the city's assumed cosmopolitanism. Towards the end of the 20th century, a new transformation of the city began, caused by rapid urbanisation and real estate development. Today, the most populous parts of Alexandria are no longer the old central districts, but the numerous new areas that have grown to the east, south, and west of the city. With few exceptions, the villas and small houses that once stood near the seafront to the east and west of central Alexandria have been demolished and replaced by high-rise buildings.

While the Alexandria of the seafront, with its Euro-cosmopolitan past, has been associated with holiday romance and images of a liberal Egypt open to the world, the Alexandria of the inland has become known as a centre of Islamist activism since the 1970s. In the past two decades, the purity-oriented Salafi movement has become a main religious player in the city, competing with the Muslim Brotherhood for followers (Decschamps-Laporte 2014). The ahistorical vision of Salafism resonates well with the drive of real estate developers to demolish and build. Notably, both Salafism and real estate speculation are truly global movements for which national borders and identities are secondary. And yet, just like rural migrants and Syrian refugees, Salafis and real estate speculators also do not fit into the standard narrative of Alexandria's cosmopolitanism. The standard seaside cosmopolitan myth is a story of past grandeur and present decline, whereby the contemporary city is not worthy of interest in its own right. Outside of Egypt, it has been reproduced by concerned journalists (Traub 2014; Hadid 2016) as well as critical intellectuals – including Edward Said (2000).

It is not only foreign visitors who are drawn to such romantic melancholia, however. At a time when both 20th-century secularist nationalist visions of a bright future as well as the material and cultural continuity of the 20th-century city began to crumble, a new interest in cosmopolitan Alexandria before the 1950s emerged. Cultural circles in the city developed an increasing interest in non-Arabic heritage. Both Egyptian state institutions and international cultural organisations and funding bodies began to actively promote a usually de-politicised, sanitised vision of past glory. This vision, which is especially, but not only, popular among liberal members of the urban bourgeoisie, is today also supported by social media pages that post photos of colonial and monarchy-era Egypt, often accompanied by explicit words of praise for a beautiful past and depreciation of the present (Ryzova 2014).

And yet, no matter how counterfactual it may be today, the idealised association of Alexandria with beauty needs to be taken seriously as something that many inhabitants of the city strive for.

Regardless of their political and religious views, their origin and their class position, inhabitants of the city generally appreciate the sea, even if they only rarely manage to take a stroll along the seafront in their free time. On warm evenings, the Corniche becomes crowded with families, couples, groups of friends, and lone strollers. Many of them sit down and look out on the sea, towards the lights that can be seen on the horizon. I have been told that those are the lights of Cyprus.

Unromantic sceptics object that it is impossible to actually see Cyprus because it is more than 500 km away from Alexandria, and that the lights on the horizon belong to ships and fishing boats. But I am not interested in

Figure 6.2 The Corniche by Ramleh Station.
Source: Photo by Samuli Schielke (2011).

questioning whether it *really* is Cyprus that they see. Instead, I am interested in the gaze itself of the night-time strollers as they look at the dim lights on the horizon. That gaze says something about the city's paradoxical location between a congested, segregated, and largely unappealing urban conglomerate stretching inland – where the vast majority of its inhabitants live and work – and the seaside as the mythological, value-laden location of the city where inhabitants and visitors can imagine and appreciate Alexandria as something beautiful and magical, even if they only rarely manage to actually go there. The Cyprus seen by those strolling on the Corniche is an intimate part of Alexandria. Rather than firmly locating Alexandria as part of the Mediterranean world, however, the gaze toward Cyprus highlights the ambiguity of the Mediterranean Sea, having become a border that divides much more so than being a means of communication that unites.

As a city mythologically located on the sea, Alexandria is also defined by the presence of borders right in its urban heart. The international border of the Mediterranean Sea is paralleled by the class boundaries that run between seaside and inland neighbourhoods, and between the many segregated suburbs even further inland. Those borders, in turn, establish the stranger or the Other as a key figure in the myths that can be told about the city. Explaining what and where Alexandria is typically involves some telling of the relationships of strangeness and alterity that structure the city and the specific spots where those relationships evolve – be they romanticised, as the 'cosmopolitan era' and the historical downtown that still embodies its memory often are, or scandalised, as the anti-city of the sub-urban crescent often is.

What makes literature interesting as a production site of such myths is that it often creates myths with a twist, myths that try to change the set-up of the stories worth telling.

Ghurbal

This is the Alexandria into which the lawyer and poet Hamdy Zidan was born in 1972: the neighbourhood of Ghurbal, south-west of the historical downtown, one of Alexandria's old 'popular districts'. His grandparents migrated to Alexandria from Upper Egypt in the first half of the 20th century and settled here. His father was a wedding singer, and Hamdy became interested in literature and music at an early age. He describes the Ghurbal of his youth as his key inspiration, a society that was at once conservative and open-minded. It is a densely populated quarter of narrow streets laid out in a modernist quadratic grid. In his childhood, the houses had only two floors built of brick and a third floor built of corrugated iron and wood. Several families shared one floor or one apartment, with a shared kitchen and bathroom. Christians, Muslims, and people from different parts of the country all lived together. With the houses fully packed, life took place in the streets. According to Hamdy, there was a magic to the streets, paved with basalt blocks, with steps of iron and stone pillars: 'This quarter creates drama and debate. The place gave me the magic key to language.'

In autumn 2011, the novelist Mukhtar Shehata and I were attending exhibitions and symposia as part of the joint fieldwork we had just begun. Among the places we frequented was El Cabina, an independent cultural space that had opened months earlier. During the five years of its existence (2011–2016), El Cabina quickly developed into a meeting point of an internationally connected, politically pro-revolution, leftist or liberal, secular cultural milieu with avant-garde tastes.

Among the events we attended was a symposium dedicated to Hamdy's poetry that took place on 17 October 2011. Connecting the memory of his youth and childhood with his free-verse poetry in the Egyptian colloquial, Hamdy outlined his literary and political vision of the city:

> I'm interested in the study of Alexandria as an example of the human con-
> dition that we live everywhere in the world – an example of pluralism,
> openness, tolerance. [...] Alexandria, starting with Alexander the Great, is
> a sentence that has no full stop, no definite end ....

The real Alexandria for Hamdy is not the bourgeois and cosmopolitan districts on the sea, but the old popular quarters housing migrants from different parts of Egypt, living together in close spaces, but feeling that the whole city is theirs. It is this city – very much the city of his childhood, still remembering the colonial era, connected to a history and looking forward to a better future, rooted yet tolerant and open-minded – that Hamdy elevated to a moral principle that must live on, despite and against the powers that have since come to dominate the city. Hamdy's vision of the city did not exclude contradictions and conflicts, but he was convinced that 20th-century popular-quarter Alexandria offered constructive solutions for coexistence in spite of its contradictions, and therefore needed to be preserved, remembered, and revived.

Locating the myth of Alexandria as an open city (in the sense of being open to the world and to difference) in its old popular quarters stands in a longer literary tradition established by Edward Kharrat (1993, 1999) and other Arabic writers (see Kararah 2006; Starr 2009; Halim 2013) before Hamdy's time. It is also reflected in Mustafa's (who is not a writer and reads little) appreciation of Bahary as the real Alexandria. According to Hala Halim (2013: 282–283), Arabic writing from Alexandria has developed the theme of Alexandrian cosmopolitanism in a decidedly different fashion than that adopted by most European writers, and at times even decidedly against it. But as Alaa Khaled (2012: 12–13) points out, even while that tradition locates cosmopolitan coexistence in popular quarters, it remains committed to the politically safe vision of a lost paradise, grounded in a sense of perpetual loss between a grim present and a golden past.

For Hamdy, a key theme in his vision is the relationship with the Other, especially the Christian and the foreigner. That relationship is at once fraught and attractive, as in his poem 'White Desire' (Zidan 2013):

Lady Faransa[3]
who collects the wax at the church
to melt it down at home to sell it back to the same Church
was our Christian neighbor.
I understood that on my own when I was little
from the large black straw cross on
the chest of her short black dress
and her silver hair the color of melted wax.
Our mute neighbor
screamed when the boiling wax spilled on her.
She screamed, and no one noticed her,
like my boiling desire within my straw heart.

I noticed a peculiar turning of tables regarding heritage and progress in these events. They took place in a cultural milieu that sees itself as progressive and in opposition to both a conservative religious current as well as the authoritarian system of the state. And yet the prevailing tone was that of a cultural critique of the forgetfulness and destruction of history involved in the Islamic revival. The self-declared cultural and literary avant-garde was upholding remembrance of the past and connectivity with tradition, in opposition to a wave of religiosity marked by a characteristically modernist oblivion in place of history.

A claim to a tradition involves a claim to power. As Talal Asad (2014) has pointed out, traditions are not taken-for-granted continuities. Rather, they are the foundation and result of struggles for power to define, reproduce, and guard them. Hamdy's myth of Alexandria as a principle of openness located in Ghurbal rather than in the seaside districts is a way to wrestle some of the power of the cosmopolitan myth from European bourgeois into Egyptian working-class hands. And yet it remains grounded in a specific, essentialised vision of what is and what is not Alexandria, and an accompanying claim by the urban intelligentsia to define the city. Although a principle of openness, Hamdy's Alexandria is not open in every direction.

A long discussion followed at the symposium. Among those asking questions was Mukhtar. He has a different relationship with the city: 'Maybe it is because I only moved to Alexandria 7 or 8 years ago, and I don't love it the same way.' He asked why the Alexandrian authors in the circle only write about the old Alexandria. He demanded a literature for and about the districts that were once small houses and gardens, and have now turned into 'a concrete jungle': al-Mandara, 45 Street, Abu Kharouf. Mukhtar claimed that these places are never mentioned in the stories of the city, but that they are even sharper and harsher places than some of the ill-famed old quarters such as Gheit El Enab (for the latter, see Abdel Meguid 1999; ElToukhi 2014). They, Mukhtar argued, are places that can and must be written about: 'Abu Kharouf can equal the Gammaliya of Naguib Mahfouz.'

Mukhtar was in this way claiming space for his own writing. At the time, he was sketching a novel (Shehata 2017) that would be located partly in Abu Kharouf and partly in Gheit El Enab.

Hamdy disagreed. He had actually lived in Abu Kharouf for more than ten years. There are writers from the bourgeois milieu who really do not know this side of Alexandria, but Hamdy knows it inside out. He argued that those suburbs are like a cancer attacking the city. In contrast to the plurality, openness, and rootedness that he identified as the characteristics of Mahfouz's Gammaliya (and his Ghurbal), Hamdy saw informal settlements like Abu Kharouf as the very opposite of the idea of Alexandria. He claimed that they are 'like Kandahar' (drawing a comparison between the Taliban stronghold in Afghanistan and the power of the Salafi movement in Abu Kharouf), a place where social relations have collapsed, embodied by adolescents from the informal areas who come to fill Ramleh Station during the Islamic feasts to harass women and anybody who looks unusual.

If for Hamdy Alexandria was a dream of a better world, a memory to revive for the sake of a better future, for Mukhtar Alexandria was the shocking reality of a divided city in which some people attempt to revive the city's cosmopolitan age, while others want to transform it into Kandahar, a seaside city in which some people living in inland informal settlements have allegedly never seen the sea. After the reading in Cabina in 2011, he criticised the downtown intellectuals for the self-imposed isolation they create by celebrating the memory of old Alexandria and rejecting the concrete jungle. In doing so, he argued to me, they close themselves up in a small circle and, by rejecting the reality of the city, fail to reach out to the concrete jungle.

Alexandria as an open city is a very likeable myth, and there are good reasons why it is reproduced by writers who hope that their city might be more accommodating towards different ways of life – especially those that are not promoted by the Islamic revival or the real estate boom. Such ways of life are indeed precarious in Alexandria outside some protected niches. However, it is not an innocent myth. It comes with a political economy. It easily legitimates the privileges of the urban bourgeoisie and the Egyptian regime. It also allows European funding agencies and visitors to leave the comfortable privileges they enjoy unchallenged. This makes it a safe (and thus also potentially profitable) topic for national projects and international cooperations alike.

As a result, the myths of the city in this tradition have become populated by two kinds of strangers: those who fit into a vision of openness, and those who must be excluded, even destroyed, in order to safeguard that openness. It is no coincidence that many (albeit not all) people who sympathised with the cosmopolitan coexistence narrative, joined forces in 2013 with militarist nationalism and either tacitly accepted or openly supported the massacring of supporters of the deposed President Morsi. With their vision of moral and confessional purity and their strong grounding in popular neighbourhoods like Abu Kharouf, the Islamists are easy to depict as the very opposite of the spirit of Alexandria as Hamdy sketched it. Seen from this point of view, they were the ideology of the anti-city, and they had to be excluded and destroyed.

The East of the City

The anti-city of the open city myth, however, actually makes up most of the city. Alexandria as I have encountered it in past years is a plural city, but not a pluralistic one. Hamdy's version of the real Alexandria as rooted in and open to difference, continuous between past and present, strikes me as more sympathetic, but I find Mukhtar's version, which foregrounds the 'concrete jungle' and the break with the past it involves, closer to the city I know.

While public sector cultural flagships like the Bibliotheca Alexandrina have engaged in nostalgic celebration of 'cosmopolitan Alexandria' (Awad and Hamouda 2006), a certain anti-nostalgic backlash has emerged in parts of the cultural scene. Interestingly, this backlash has been produced partly by the very same people who, 20 years earlier, spearheaded the search to reconnect their urban present with its past non-Arabic inhabitants and literatures (e.g. Raouf 2016). Paradoxically enough, it is being articulated by people who are internationally well-connected and who read both English and Arabic literature and social theory (and some read French) – that is, people who would easily qualify as cosmopolitan by most counts.

Among them is Ali Al-Adawy, born in 1985 in the eastern suburb of Abu Qir, organiser of film and cultural programmes, writer and editor. Since 2014, he and some of his friends have been working to put together a research and film project about the East of the City (Sharq al-madina) which, in their view, has replaced the historical downtown of al-Manshiya and Ramleh Station as the centre of the city. The East of the City – especially the district of Sidi Bishr – represents an anonymous, consumerist, at once conservative and individualist form of urbanity influenced by Egyptian migration to the Gulf, the import-export business, the Islamic revival, and unrestrained real estate expansion. If the old central districts stand for what Alexandria may once have been, Sidi Bishr shows what it is now becoming, and quite literally so: the race to demolish villas and smaller apartment buildings and to build 15-storey high-rises in their place began in the East of the City around the turn of the millennium.

Since 2011, this demolition and construction boom has engulfed almost the entire city. Old popular quarters like Bahary, Ghurbal, and others have been thoroughly transformed, a large portion of their older houses replaced by high-rises. Exclusive new projects on land fill are making the sea inaccessible to the general public and no longer visible from many parts of the Corniche (El Nemr 2017). Beaches have already charged for admission for some time, following two decades of privatisation of public space. Until now, however, viewing the sea had still been free of charge.

Some urban activists try to document and protect urban architectural heritage. But with government and private interests aligned towards generating maximum profit from construction and real estate, a progressive erasure of the city appears unstoppable. And with the gradual disappearance of the sea shore behind resorts on land fill, Alexandria may one day no longer be a city by the sea for its non-privileged inhabitants.

In search of ways to overcome what he sees as an unproductive nostalgia in writing about the city, Ali turned to the work of Walter Benjamin. With funding from the Goethe-Institut he organised a workshop about 'Benjamin and the City'. Ali hoped that Benjamin's way of writing about Berlin and Paris (Benjamin 1991 [1939], 2019) might provide guidance for overcoming the cosmopolitan nostalgia, and for deconstructing any and all narratives of the city:

> The idea of the narrative of the city – be it an old and conservative narrative, or a contemporary one – is an ideological notion that constantly relies on the historical, political, social, and economic framework and context. It expresses the reality that it in a way produced despite all its attempts to disguise it.
>
> (Al-Adawy et al. 2016: 6)

The main outcome of the workshop was a small collection of essays that was presented in El Cabina on 10 March 2016. It was something more contradictory than what Ali may have been aiming for. The texts were evenly divided between two approaches: Abdelrehim Youssef, a teacher, poet, and cultural programmer at El Cabina, and Yasmine Hussein, a researcher at the Alexandria Library and photographer, had each written childhood memories with an eye for minute details and personal experiences, inspired by 'the dominance of the poetic' (in the words of Abdelrehim) in Benjamin's *Berlin Childhood around 1900* (Benjamin 2019). Hager Saleh, an M.A. student in history, and Hakim AbdelNaim, an actor and theatre director, produced more comprehensive critical engagements with the city. An expression that came up in the latter two texts was *al-madina al-za'ila* 'the perishing/non-permanent city', a vision of a city in a constant process of erasure. In the words of Hager Saleh:

> Thus, the city likes to show off its passing/perishing (*za'ila*) cosmopolitan-ism. It hides its history and covers it with dust as if it were a disgrace that deserves to be erased, and then again boasts of it with insolence. The city persistently reinvents itself, carrying a new face in every era and hiding its old face under rubble.
>
> (Saleh 2016: 10)

A long discussion followed the presentation. Although it had not been a major theme of the workshop, a controversy about 'nostalgia or not' dominated the discussion.

The theatre director and manager of a performing arts NGO Ahmed Saleh claimed: 'Also today's writings were loaded with nostalgia, just like the writings of the past 20 years. What new does Benjamin offer?' Abdelrehim disagreed and pointed out that three of the five texts presented were critical of nostalgia; only his and Yasmine's leaned towards nostalgia. Hakim commented that Ahmed probably intentionally played the role of the provocateur. More important than nostalgia or not, he went on, was to question the classist aspirations of the specific nostalgia for a city by the sea. From what kind of societal configuration did that city emerge?

The poet and guest participant in the workshop Ahmed Abdel Gabbar defended a nostalgic relatedness to the past and its traces:

> Cavafy also didn't write of the Alexandria of his age, but of the Hellenic era. That history is still present, under the earth. Kharrat's popular quarters and Durrell's Cecil Hotel are still there in the city you move in. While I speak I see the ruins of the demolished Rialto Cinema. But it was there. Even if only in the layout of the streets, the traces remain with us. I see nostalgia positively, if it means that I know what I write about.

Hager countered: 'We are drawn to cosmopolitan longing because of its dramatic touch. Like classical tragedy, it is attractive.' Addressing the many historical periods of the city and its varying centres and dominant groups, she pointed out that the location of the city itself was constantly on the move: 'The city is not something solid.'

Mohamed Elshahed, editor of the *Cairobserver* magazine on urbanity and architecture in Egypt, insisted on a more complex picture. The way we speak about the past reflects the way we speak about the present and reproduces its blind spots, he argued. What is left out in the binary of the *khawagas* (as in Durrell) and the popular quarters (as in Kharrat), he argued, is the social history of Alexandria from the 1940s to 1960s, a period of major social mobility for urban inhabitants of Egyptian origin, when many rural migrants climbed into bourgeois society.

Abdelaziz ElSebaei, one of the founders of the Eskenderella Association who had left it in 2013 along with Maher Sherif, intervened to problematise what he called 'the passion for the city':

> It has become a sort of national disease. I'm not against engagement with the city. But we always try to reach back to times before us. Myself, I'm not as upset today as I was twenty years ago when an old house is demolished.

The presentation of the 'Benjamin and the City' workshop in March 2016 marked a departure from the nostalgic tone that had dominated the poetry symposium in 2011. Svetlana Boym (2001) argues that nostalgia may come in restorative and reflective varieties; Ahmed Abdel Gabbar clearly claimed the contemporary usefulness of a reflective nostalgia in his comment. The critique of cosmopolitan nostalgia, however, equates it with a futile attempt to restore something that can no longer be retrieved, and probably was never really so beautiful in the first place. Such critique is in line with an emerging shift from the binary towards the fragmentary in writings about the city, such as in Alaa Khaled's *Alexandrian Faces* (2012). The Benjamin workshop also coincided with other cultural events and publications in 2016 (see, e.g., Nizar 2016) that balanced a reflective-nostalgic search for ways to remain connected to the city's 20th-century history and the positive values it might represent on the one hand, with a demand to recognise the self-erasing, conflicted, and divided character of the city's present and past, on the other.

What had changed? A generational shift is part of the story. Some participants in the Benjamin workshop, notably Hager Saleh and Hakim AbdelNaim, are young enough to have experienced their generational formation during the revolutionary period. But others had been active in the scene even before 2011; Abdelaziz ElSebaei was born in 1949. Nostalgia in all varieties is a reflection of the present against which it is posited, and the present had changed. For those who in 2016 questioned the nostalgia for old Alexandria, the very recent events of the revolution provided a more pertinent nostalgic relation to the present. Theirs was now a more conflicted and combative longing for a future very recently lost, and the myth of an unchanging spirit of a true Alexandria appeared less helpful to provide orientation in the city and country in which they lived. By 2016, after a defeated revolution and a victorious construction boom, the topos of unsolved conflicts and permanent erasure had become more pertinent, and the nostalgic vision of connectedness and openness more difficult to maintain (see also Faruq 2017).

In a short text published a year later, Hakim AbdelNaim made explicit the link between his suspicion towards nostalgia and the trauma of the defeated revolution:

> All places are accompanied by trauma, by post-traumatic stress disorder, by an enormous affective experience that was not completed, that found no occasion to have a light ending, or even a heavy one but without a sudden cut, as if a person dies burning and remains in his final state, state of trauma … and who knows if he died of trauma or of heat? I detest longing and everything that has a relation with longing and everything that makes me feel that it is part of the longing I detest. I fear it and its closed circle.
>
> (AbdelNaim 2017)

And yet by the very virtue of their intensive concern with the city as such, pieces of writing that aren't nostalgic also contribute to the mythologisation of Alexandria. It is a different myth, however. It tells of Alexandria as a perishing, non-permanent city of conflicts, fragmentation, and erasure.

The myth of the non-permanent city has the paradoxical advantage over the myth of the cosmopolitan open city that it is more inclusive. It has space for both al-Manshiya and al-Mandara, both Bahary and the Chinese Housing. The myth of the non-permanent city is cosmopolitan in its own way, in the sense that it tells of the urban coexistence of difference. However, it highlights conflicts over harmony. I am definitely not impartial in this matter. Part of a wider shift in academic interest towards understanding Alexandria as an ordinary city in the present (El Chazli 2018; Jyrkiäinen 2019), this essay contributes to the narrative that highlights conflicts and erasure. With the publication of the Arabic version of this essay in December 2016, it became a part of the conversation it addresses (see Schielke 2016). And yet the issue at hand is not an opposition between a romantic fantasy of what Alexandria might

Figure 6.3 Inland al-Mandara in the East of the City.
Source: Photo by Samuli Schielke (2016).

once have been and a realistic recognition of what the city really is. The very question of what or where the city 'really' is, is an exercise in fantasy.

Every claim to have located the city is the product of a certain politically and morally loaded work of imagination (Chiti 2016). Hamdy's rewriting of the cosmopolitan myth from the point of view of Ghurbal, Mukhtar's emphasis on the 'concrete jungle' of Abu Kharouf, even Ali's search to deconstruct the narrative unilinearity of 'the city', are all expressions and draft blueprints of specific urban mythologies where neighbourhoods, streets, and fictional characters embody specific affective, political, and moral visions and conflicts. Mustafa's appreciation of Bahary as the real Alexandria and the Chinese Housing as his America shows that these mythologies can coexist, and find resonance even beyond dedicated literary circles.

Taking myths seriously as social theories means considering the possibility that some of them may provide a more likely true and helpful account of the realities they describe than others do. The essentialising utopias of an organic, true, better city that are evoked by cosmopolitan nostalgia in its popular-quarter and seaside varieties alike need to be recognised for what they are: dreams of and strivings for beauty and ease of life, made only more compelling by their increasingly counterfactual character. However, anti-utopian myths of erasure and conflicts can provide a better orientation for understanding what kind of a city Alexandria is today, where it is, and in what directions it is moving.

Notes

1 Acknowledgements: In addition to those mentioned by name in the essay, thanks are due to Bettina Gräf, Aliaa ElGready, Abeer Hosni, Michael Lambek, Amany Maher, Ayşe Öncü, Amr El Sabbagh, Haitham Shater, Daniela Swarowsky, Jelena Tošić, Jessica Winegar, the anonymous reviewers of *HAU* journal (who read a version of this essay that was not published), the Gudran Association, the Bibliotheca Alexandrina, and the Trajectories of Lives and Knowledge research group at Leibniz-Zentrum Moderner Orient. Early versions of this essay were presented in talks at Sabanci University in Istanbul, 2012; Northwestern University in Evanston, IL, 2013; the conference of the German Orientalist Society in Münster, 2013; Zentrum Moderner Orient in Berlin, 2013, 2016, and 2018; the biannual meeting of the German Anthropological Association in Marburg, 2015; and the University of Heidelberg, 2017. Research for this article was made possible by funding from the German Federal Ministry of Education and Research for the junior research group In Search of Europe: Considering the Possible in Africa and the Middle East.
2 While many historians tend to avoid the Euro-centric pitfall, recent historiography of Ottoman cities (e.g. Freitag and Lafi 2014) also tends to equate cosmopolitanism with the well-ordered urban coexistence of different ethnic and religious groups, and not with border-crossing lives or a sense of worldliness that would amount to one sort or another of a 'citizenship of the world'.
3 Faransa, Arabic for France, was common as a Christian girl's name in the early 20th century.

References

Abdel Gabbar, Ahmed. 2013. 'Al-'Agami: al-'umran la ya'ti gharban.' *Tara al-Bahr* 1 June. https://tinyurl.com/yy2mqn3o.

Abdel Meguid, Ibrahim. 1999. *No One Sleeps in Alexandria*. Translated by Farouk Abdel Wahab. Cairo: The American University in Cairo Press.

Abdel Meguid, Ibrahim. 2005. *Birds of Amber*. Translated by Farouk Abdel Wahab. Cairo: The American University in Cairo Press.

Abdel Meguid, Ibrahim. 2013. *Al-Iskandariya fi ghayma*. Cairo: Shorouk.

AbdelNaim, Hakim. 2017. 'Mudun al-sadma.' *Tara al-Bahr* 4: 10.

Al-Adawy, Ali Hussein, Hakim AbdelNaim, Hager Saleh, Abdelrehim Youssef, Kholoud Saeed, and Yasmin Hussein. 2016. *Benyamin wa al-madina: Hal yumkin sard madina? Nusus*. Alexandria: Yadawiya. https://tinyurl.com/yxg7x5zn.

AlSayyad, Nezar, and Ananya Roy. 2005. *Urban Informality: Transnational Perspectives from the Middle East, Latin America, and South Asia*. Lanham, MD: Lexington Books.

Asad, Talal. 2014. 'Thinking About Tradition, Religion, and Politics in Egypt Today.' *Critical Enquiry*. http://criticalinquiry.uchicago.edu/thinking_about_tradition_religion_and_politics_in_egypt_today/.

Awad, Mohamed, and Sahar Hamouda, eds. 2006. *Voices From Cosmopolitan Alexandria*, Volume 1. Alexandria: Bibliotheca Alexandrina.

Barthes, Roland. 1970. *Mythologies*. Paris: Éditions de Seuil.

Benjamin, Walter. 1991 [1939]. 'Paris, Capitale du XIXeme Siècle.' In Benjamin, Walter, *Gesammelte Schriften, Band 5: Das Passagen-Werk*, edited by Rolf Tiedemann, 60–77. Frankfurt a.M.: Suhrkamp.

Benjamin, Walter. 2019. *Werke und Nachlaß: Kritische Gesamtausgabe - Band 11: Berliner Chronik/Berliner Kindheit um neunzehnhundert*, edited by Burkhardt Lindner and Nadine Werner. Frankfurt: Suhrkamp.

Bissell, William Cunningham. 2005. 'Engaging Colonial Nostalgia.' *Cultural Anthropology* 20 (2): 215–248.

Boym, Svelana. 2001. *The Future of Nostalgia*. New York: Basic Books.

Chiti, Elena. 2016. 'Quelles marges pour quels centres? Perceptions arabes et européennes d'Alexandrie après 1882.' In *Etudier en liberté les mondes méditerranéens*, edited by Leyla Dakhli and Vincent Lemire, 491–501. Paris: Publications de la Sorbonne.

Decschamps-Laporte, Laurence. 2014. 'From the Mosque to the Polls: The Emergence of the Al Nour Party in Post-Arab Spring Egypt.' *New Middle Eastern Studies* 4. www.brismes.ac.uk/nmes/archives/1350.

El Chazli, Youssef, ed. 2018. 'Everyday Alexandrias: Plural Experiences of a Mythologised City.' Special issue in *Égypte/Monde Arabe* 17.

El Nemr, Nahla. 2017. 'Alexandria's Vanishing Sea Shore.' *Zenith* 28 June. Accessed 28 January 2018. https://magazine.zenith.me/en/society/identity-and-corruption-alexandria.

ElToukhi, Nael. 2014. *The Women of Karantina: A Novel*. Translated by Robin Moger. Cairo: The American University in Cairo Press.

Fahmy, Khaled. 2006. 'Towards a Social History of Modern Alexandria.' In *Alexandria: Real and Imagined*, edited by Anthony Hirst and Michael Silk, 281–306. Cairo: American University in Cairo Press.

Faruq, Osama. 2017. 'Wahm al-kuzmubulitaniya fi Bayrut wa l-Iskandariya wa Azmir.' *Al-Mudun*, December 24. Accessed 28 January 2018. https://tinyurl.com/y5x7s5g8.

Freitag, Ulrike, and Nora Lafi, eds. 2014. *Urban Governance under the Ottomans: Between Cosmopolitanism and Conflict*. London: Routledge.

Hadid, Diaa. 2016. 'Remembering My Mother's Alexandria.' *New York Times*, November 29. Accessed 28 February 2019. https://nyti.ms/2geuApy.

Halim, Hala. 2013. *Alexandrian Cosmopolitanism: An Archive*. New York: Fordham University Press.

Hanley, Will. 2017. *Identifying with Nationality. Europeans, Ottomans, and Egyptians in Alexandria*. New York: Columbia University Press.

Hazem, Menatallah. 2006. *Modern & Contemporary Alexandria in Fiction: A Bibliography*. Alexandria: Bibliotheca Alexandrina. www.bibalex.org/libraries/presentation/static/Modern_Contemporary_Alexandria_in_Fiction.pdf.

Jyrkiäinen, Senni. 2019. *Virtual and Urban Intimacies: Youth, Desires and Mediated Relationships in an Egyptian City*. PhD diss., University of Helsinki.

Kararah, Azza. 2006. 'Egyptian Literary Images of Alexandria.' In *Alexandria: Real and Imagined*, edited by Anthony Hirst and Michael Silk, 307–321. Cairo: American University in Cairo Press.

Kazamias, Alexander. 2009. 'The "Purge of the Greeks" from Nasserite Egypt: Myths and Realities.' *Journal of the Hellenic Diaspora* 35 (2): 13–34.

Khaled, Alaa. 2012. *Wujuh sakandariya: Sirat madina.* With photographs by Salwa Rashad. Cairo: Shorouk.

Kharrat, Edward. 1993. *Girls of Alexandria.* Translated by Frances Liardet. London: Quartet Books.

Kharrat, Edward. 1999. 'Interview with Edward Kharrat.' *Banipal* 6. www.banipal.co.uk/selections/51/145/edwar-al-kharrat/.

Lévi-Strauss, Claude. 1955. 'The Structural Study of Myth.' *The Journal of American Folklore* 68: 428–444.

Manoukian, Setrag. 2012. *City of Knowledge in Twentieth Century Iran.* London and New York: Routledge.

Newcomb, Rachel. 2017. *Everyday Life in Global Morocco.* Bloomington: Indiana University Press.

Nizar, Nermin. 2016. 'We Came to Alexandria Looking for ....' *Mada Masr*, 12 May. Accessed 26 December 2019. https://madamasr.com/en/2016/05/12/opinion/u/we-came-to-alexandria-looking-for/.

Raouf, Khaled. 2016. 'Da' al-hanin: Da' al-"Hamada".' *Tara al-Bahr* 1: 30–31. https://tinyurl.com/y68776x3.

Robinson, Jennifer. 2006. *Ordinary Cities: Between Modernity and Development.* London: Routledge.

Ryzova, Lucie. 2014. 'Nostalgia for the Modern: Archive Fever in Egypt in the Age of Post-Photography.' In *Photographic Archives and the Idea of Nation*, edited by Costanza Caraffa and Tiziana Serena, 301–318. Berlin and Munich: De Gruyter.

Said, Edward W. 2000. 'Cairo and Alexandria.' In *Reflections on Exile and Other Essays*, edited by Edward W. Said, 337–345. Cambridge, MA: Harvard University Press.

Saleh, Hager. 2016. 'An tata' madina tajtham 'ala atlal mudun qadima.' In *Benyamin wa al-madina: Hal yumkin sard madina? Nusus*, edited by Ali Hussein Al-Adawy et al., 9–13. Alexandria: Yadawiya. https://tinyurl.com/y37qt59m.

Schielke, Samuli. 2016. 'Ayna taqa' al-Iskandariya? Asatir al-madina wa al-madina al-naqid fi al-'asr ma ba'd al-kuzmubulitaniya.' Translated by Abdelrehim Youssef. *Tara al-Bahr* 3: 12–27. Accessed 28 January 2018. https://tinyurl.com/yyxe34pm.

Schielke, Samuli, and Mukhtar Saad Shehata. 2016. 'The Writing of Lives: An Ethnography of Writers and Their Milieus in Alexandria.' *ZMO Working Papers* 17. Accessed 28 January 2018. http://d-nb.info/1122236654/34.

Shehata, Mukhtar Saad. 2017. *Asafra Qibli: Riwaya.* Cairo: Battanah.

Simone, AbdouMaliq, and Edgar Pieterse. 2017. *New Urban Worlds: Inhabiting Dissonant Times.* Cambridge: Polity Press.

Starr, Deborah. 2009. *Remembering Cosmopolitan Egypt: Literature, Culture, and Empire.* London: Routledge.

Traub, James. 2014. 'The Lighthouse Dims.' *Foreign Policy*, 23 December. http://foreignpolicy.com/2014/12/23/lighthouse-dims-egypt-alexandria-salafists-mubarak-sisi-longform/.

Zidan, Hamdy. 2013. 'Raghba bida/White Desire.' Translated by Jennifer Peterson. In *Iskenderiyan Standards: Nusus*, edited by Maher Sherif. Alexandria: Yadawiya.

Zubaida, Sami. 2011. *Beyond Islam: A New Understanding of the Middle East.* London and New York: I.B. Tauris.

7 Jerusalem's lost heart

The rise and fall of the late Ottoman city centre

Yair Wallach

Jerusalem is one of the prime cases of 'divided cities' discussed in urban studies, alongside Belfast, Beirut, and Nicosia (Calame et al., 2009). Such cities are characterised by sharp residential segregation according to ethnic, national, or religious identity. In Jerusalem, effectively all neighbourhoods are understood as either 'Jewish' or 'Arab', and residential segregation is almost total.[1] The absence of a shared city centre is another typical characteristic of divided cities. Jerusalem is served by two distinct business centres: the Jewish-Israeli town centre of Jaffa Road, and the Palestinian business district of Damascus Gate. This division is the legacy of 19 years (1948–1967) in which the city was physically divided into Israeli West Jerusalem and Jordanian East Jerusalem. Physical partition ended in 1967, yet the division of commercial centres has endured since, despite – and perhaps due to – the heavy-handed planning and construction interventions of Israeli 'unification'. The persistence of the division points to the limits of Israel's unilateral annexation project. The absence of a shared city centre is tightly connected to the division of the city's neighbourhoods according to an ethno-national logic. The absence of a common civic ethos makes it difficult if not impossible to sustain a city centre that could claim to serve all the city's populations.

But this was not always the case. From the 1880s to the 1930s, Jerusalem had a modern city centre offering civic, cultural, and commercial amenities to Jerusalem's diverse constituencies. It was located in Jaffa Gate, which today offers liminal centrality. The area began to develop in the 1880s, and by 1900 Jaffa Gate was the undisputed heart of the city. Its character was decidedly Ottoman, cosmopolitan, and non-sectarian. But British colonial rulers, who occupied the city in 1917, saw no merit in this area. British policy makers resolved to destroy much of the modern Jaffa Gate quarter in order to separate the Old City from the new parts. The Ottoman vision of a civic, non-sectarian and modern Jerusalem – embodied in the Jaffa Gate area – was anathema to British officials who saw Jerusalem as an ancient city and a patchwork of ethnic and religious congregations, each in their own neighbourhoods. The British were unable to fully implement their plans, but their policies during 30 years of rule removed civic activity from the Ottoman centre to other parts of the city, as Jerusalem became increasingly segregated between Jews and Arabs. In late 1947, the former centre became a battleground between Zionist militias and Arab nationalists, and in the aftermath of the 1948 war, a no man's land between Israel and Jordan. After the occupation of East Jerusalem in 1967, Israeli planners finally carried out the British colonial vision and physically destroyed almost the entire area. This staggered process of destruction was paralleled by an almost complete erasure of the Ottoman city centre from cultural memory and literature on Jerusalem. The civic and political importance of the area has been systematically downplayed or ignored altogether. Present-day Jerusalemites not only have no shared city centre: they are also not aware that Jerusalem ever had such a centre.

A city centre is typically not a residential neighbourhood, yet its function and character are crucial to the configuration of the city's neighbourhoods. As the primary node in the city's networks of power, movement, and exchange, the city centre encodes, in a physical way, the city's ethos and material experience, its hierarchies, governing discourse, and logic of commerce. Neighbourhoods are often defined by their proximity and connection to the centre; given the role of the centre as a transport hub, connections between neighbourhoods are mediated through it. Local neighbourhood shopping facilities and markets are often identified by their relation to the main shopping district. And above all, the centre embodies – in architecture, symbols, signs, and amenities – the dominant civic ethos, political order, and hegemonic understanding of the city, which affect and to a large degree define all the neighbourhoods of the city. The hegemonic civic discourse is never neutral or natural; it is unavoidably political and articulated against state and global frameworks; it is inclusive of some groups, and exclusive of others. Transforming the city centre – or in Jerusalem's case, destroying it altogether – can have an inevitable effect on the city's neighbourhoods, which change as the meaning of the entire urban configuration shifts. Once a city centre is destroyed, it is inevitably replaced by a new centre or centres, as happened in Jerusalem after the 1948 war. But these embody a different urban order, and a different urban economy, as the city is forced to adjust its layout and logic. In Jerusalem, the destruction of the city centre led to a radical transformation of its urban space that entrenched and normalised urban segregation as an inevitable trait of the city.

This chapter investigates the emergence of Jerusalem's late Ottoman city centre and its demise and subsequent destruction from the late nineteenth century to the late twentieth century. It seeks, firstly, to make a claim for the existence of such a civic space, its embeddedness in Ottoman reforms and modernisation, and the manner in which it figured in Jerusalem's neighbourhood configuration. It discusses the erasure of the city centre from the historiography of Jerusalem and from cultural memory. It then interrogates the staggered demise and destruction of the site, from British planning measures, through the 1948 war, to post-1967 Israeli demolitions and transformation of the area as part of a larger refashioning of Jerusalem as a segregated and divided city.

The late Ottoman city centre

Throughout most of the nineteenth century, Jerusalem's political and commercial centre was located within the inner parts of the walled city, as it had been for many centuries. The key bodies of local government – the Sarai (governor's palace), the Municipality (established in the 1860s), and the Islamic court – were all located in the vicinity of the Haram al-Sharif. Jaffa Gate was the only gate in the western side of the city walls. Its Arabic name was *Bab al-Khalil*, Hebron Gate, as it was the point of departure of the road leading south to Bethlehem and Hebron. It was also the starting point of the road westwards to Jaffa, and its European name 'Jaffa Gate' (in Hebrew, *Sha'ar Yafo*) reflected the growing importance of connections to Jaffa as Palestine's main port and gateway for tourists and trade. Like all other city gates, it was kept locked after dark. There were hardly any buildings outside the walls in that area. Those arriving at the gate noted its stark and solemn appearance, in contrast with the bustle of the city's inner streets. 'I remember well the moment of our arrival in Jerusalem in front of the Gate of Bab al-Khalil' wrote the Jaffa-born Yoseph Eliyahu Chelouche of his visit to Jerusalem as a small child in 1876. 'The sight of the city walls terrified and upset my young soul and it felt we were entering a sealed and closed city.' That stark impression of the gate contrasted sharply with the vivid activity inside the walls, where 'the city was bustling with people, and all the roads and alleyways were full of men, women, donkeys and sheep' (Chelouche, 2005, p. 25).

However, from the 1880s onwards, shops, hotels, banks, and other institutions were constructed outside Jaffa Gate, in immediate proximity to the city walls, along both sides of the road leading to Jaffa, and on the street leading to the Mamilla Islamic cemetery. The development outside the walls coincided with new buildings and changes inside the walls, from Jaffa Gate's plaza – where hotels, souvenir shops, tourist agencies, and cafés were opened – through the revamped Batrak Market, to the new Muristan's Aftimus (Euthymius) Market, developed between the 1880s and 1903 (Ben-Arieh, 1984, pp. 225–226). The result of this development within and outside of the walls was a contiguous area of commercial activity, extending from the western part of the walled city through the open plaza of Jaffa Gate and along Jaffa Road. The gate was no longer kept locked at night, and in 1898 it was rendered unnecessary by the filling of the moat, allowing for a large opening with free and easy access for traffic. The opening, created by the Ottoman authorities before Kaiser Wilhelm II's visit to Jerusalem, conveyed in clear and visible terms the Ottomans' resolve to open Jerusalem to modernity. The commercial life of the city moved from the inner markets of the Old City to the new urban district. Parts of the sixteenth-century Ottoman city walls disappeared from view within an urban sprawl that created continuity, and no visual separation, between old and new.

The area hosted the central Ottoman post and telegraph office, as well as the postal services of France, Austria-Hungary, and Italy. Bank branches included Credit Lyonnais, Imperial Ottoman Bank, Anglo-Palestine Company, the Deutsche-Palästina Bank, Palestine Commercial Bank as well as the banks of Jerusalem-based financiers Valero, Hamburger, and Frutiger (Ben-Arieh, 1986, pp. 378–385; Glass and Kark, 2018). It was a major hub for tourism, with several large hotels, such as the Fast Hotel, Imperial New Hotel, and Kaminitz Hotel (Chapman, 2018). Thomas Cook tourist agents had their office in this vicinity, and there were several souvenir shops. Photo studios – the quintessential trade of fin-de-siècle progress – proudly announced their services along both sides of Jaffa Road (Nassar, 2003; Sheehi, 2015). Local shops prided themselves on their imported European goods, such as food products, alcoholic liqueurs, clothes, clocks, and gramophones, marking a clear difference to the traditional markets inside the city walls.

'Jaffa Gate is now the biggest centre of our city, in terms of people and carriages passing through it' wrote one resident in a local paper in 1905.[2] Jaffa Gate was the main local coach terminal, which served the train station and the southern neighbourhoods, as well as the north-western neighbourhoods such as Mea Shearim and the village of Lifta, soon to be incorporated into the city. As Jerusalem was spreading rapidly in geographical terms, these connections became crucial – it was no longer possible to navigate the city only by foot. Ottoman authorities planned to make Jaffa Gate into the central hub for the Jerusalem tramway network – with three lines extending north-west and south (Dimitriadis, 2018).

The Jerusalem Municipality, the most important local political organisation, moved to Jaffa Road in 1896 (Tsoar and Aaronsohn, 2006). A short walk from the gate, up Jaffa Road, were the municipal gardens, with a café and frequent performances by the local Ottoman military band. A new water fountain was constructed just outside the gate in 1900. Most notable was the clock tower (Figure 7.1), erected in 1908 on top of the gate to celebrate 33 years of the reign of Sultan Abdulhamid II (Lemire, 2017). Similar clock towers were erected in city centres throughout the Ottoman Empire signalling the introduction of Western notions of time and public space (Wishnitzer, 2015). The clock tower of Jerusalem was funded by local donations and was a source of civic pride for the local population, as we find in the writing of local Christian Arabs and Sephardic Jews (Jawhariyyeh, 2013; Yehoshua, 1981, pp. 24–25). Other civic institutions included the Chamber of Commerce, a public theatre, and cafés. The offices of the local Arabic newspapers *al-Quds* and *al-Asma'i* were also found here. The area was the first in Jerusalem to be lit at night, and to be cleaned and washed regularly by the municipality (Ben-Arieh, 1986, p. 359).

The development of Jaffa Gate area has long been noted by historians, who have highlighted many of the details discussed above. And yet the civic significance of the area, and the implications of this transformation for Jerusalem as an urban configuration, have largely been ignored. The historiography did not identify the site as the 'city centre'; indeed, the term 'Jerusalem's late Ottoman city centre' is not used. Strikingly, on this point, there has been little difference between Zionist scholars and scholars who are more sympathetic to the Palestinian perspective. Academic scholarship and general audience publications on Jerusalem fail to acknowledge the emergence or the destruction of a city centre. The municipality, clock tower, and the new water fountain are often mentioned either in detail or in passing, but the accumulation of these aspects into a new urban and civic configuration is not acknowledged. This omission contributes to the widespread perception of Jerusalem as a city that has always been characterised by division and segregation.

Figure 7.1 Celebrations of the completion of the clock tower, 1907. On the right, the Sabil, built in 1901.

Source: Israeli National Photograph Collection. Public Domain.

Detailed discussion of the development of Jerusalem in the nineteenth and twentieth centuries can be found in the rich body of literature by Israeli historical geographers, most notably Yehoshua Ben-Arieh and Ruth Kark as well as architectural and urban historian David Kroyanker. These scholars, whose work has been published since the 1970s, have produced an impressive body of knowledge. They have relied primarily on European and Hebrew sources, and have largely ignored Arabic and Ottoman sources. In these accounts, the development of Jaffa Gate appears as an incremental part of the expansion of the city in terms of population, built-up area, and commercial activity (Ben-Arieh, 1984, 1986; Kark and Oren-Nordheim, 2001; Kroyanker, 2005, 2009). They emphasise the commercial importance of Jaffa Gate, and sometimes refer to it as Ottoman Jerusalem's Central Business District (CBD). But they do not take into account the site's political meaning and its civic role. In the most detailed study of the development of Jaffa Road between 1860 and 1948 (Tsoar and Aaronsohn, 2006) the authors state that the Jaffa Gate area 'functioned as a mixed CBD: Muslim, Christian and Jewish'. The use of the term 'mixed', rather than 'public' or 'common', is not accidental here, but indicates the dominant view of Jerusalem as essentially divided and segregated along ethno-religious lines. The city is seen as composed of distinct religious communities, which can come together, at most, for the purposes of commercial enterprise and exchange. In line with accounts by Western scholars and visitors to Jerusalem since the nineteenth century, this scholarship has presented the walled city as composed of four clearly marked ethno-religious quarters (Armenian, Jewish, Muslim, and Orthodox-Christian). The familiar pattern of the four quarters divides the cross-shaped walled city into clearly demarcated and separate ethno-religious spaces. In a city where confessional identity was so dominant, and where space always 'belonged' to one group, an overriding common identity was non-existent or extremely weak. 'Public' space did not exist, and could not have existed. Commercial areas, such as Jaffa Gate or the markets within the Old City, were, in this reading, no more than neutral spaces, in which people from different quarters 'mixed', but did not engage as members of the same community (Kark and Oren-Nordheim, 2001, p. 60). This view of Jerusalem as essentially segregated and dominated by religious identity, corresponds to a notion of Ottoman Jerusalem as inherently 'backward', 'colourful', and 'Oriental' (Kroyanker, 2005). The development of Jerusalem is usually understood as driven primarily by foreign elements: Western imperial powers, Jewish philanthropic organisations and Zionism, which brought modernity and dynamism to an Ottoman backwater. This narrative is also echoed in popular histories of the city (Gilbert, 1996; Montefiore, 2012).

Such views have not gone unchallenged. The magnificent two-volume *Ottoman Jerusalem* presents a profoundly different understanding of the city's early modern and modern history (Auld and Hillenbrand, 2000). The book explores the Islamic, Arab, and Ottoman aspects of Jerusalem between the sixteenth and twentieth centuries as a living city full of culture. The compendium has 36 essays on all aspects of life in the city, from the Dome of the Rock to libraries. But it has no mention or discussion of the modern Ottoman city centre. The development of the Jaffa Gate area is only mentioned in passing, as 'the busiest commercial artery of the city' with hotels, souvenir shops, and coffee houses (Auld and Hillenbrand, 2000, pp. 260–262). The emergence of the city centre is also not mentioned in the chapter on late Ottoman Jerusalem in the volume *Jerusalem in History*, edited by renowned Jerusalem scholar Kamil al-'Asali (Schölch, 2002).

Ottomanism, its possibilities and contestations

The last two decades have seen the emergence of a new historiography of late Ottoman Jerusalem (and Palestine more generally) that has sought to approach the late Ottoman era in a new way. Rather than project the Israeli–Palestinian conflict backwards onto the nineteenth century, this literature explores the possibilities of

Ottoman modernity. Ottoman citizenship is seen as an overarching multi-confessional framework that allowed local agents to take an active role in civic development. Rather than exceptionalise Jerusalem, this literature has aimed to position it within the broader context of modernising urban centres of the Ottoman Empire, which has seen growing interest (Hanssen et al., 2002; Freitag and Lafi, 2014). This literature has provided a robust challenge to the view of the city as essentially segregated and divided. Salim Tamari has shown that the paradigm of Jerusalem's Old City's four ethno-religious quarters was a European interpretation of local geography that did not correspond to the spatial perceptions of local residents or the Ottoman authorities. Jerusalem's urban layout was composed of a very different arrangement of neighbourhoods, which were far from religiously homogenous (Tamari, 2009). This new reading of the city's social-spatial units allows us to view an entirely different configuration of the late Ottoman city, which has implications for our reading of the city centre. Tamari's later work has explored in a rich and nuanced manner the power of Ottomanism (Osmanlıcılık) as an ideology and organising framework in late Ottoman Jerusalem, not only as a cross-confessional national identity but also one of modernity and progress, at a time of rapid change. At the same time, Ottomanism was a project open to interpretations and contestations, as notions of identity, society, and space shifted (Tamari, 2017).

The new literature's contribution has been in studying the late Ottoman period not as a prelude to the conflict between Zionism and the Arab national movement, but rather as an era of dramatic change of its own logic and discourse (Büssow, 2011). Ottoman citizenship is seen as constitutive of a national identity that could accommodate significant ethnic, confessional, and linguistic differences, allowing for new forms of politics, ideas, and expression (Campos, 2010; Jacobson, 2011). Michelle Campos has argued that Ottoman governance reforms of the 1860s, and local bodies such as the municipality and the district councils, engendered a sense of urban citizenship that became more vocal after the 1908 constitutional revolution. Local (multi-confessional) elites actively discussed infrastructure projects, commercial development, and political freedoms. Campos's account of 'shared urban space' does not address the built environment and physical urban space, but rather focuses on more abstract 'spaces' such as the press, and commercial and civic institutions, such as the Chamber of Commerce and Freemasons' lodges.

Abigail Jacobson was the first scholar to identify the civic significance of the Jaffa Gate area in the late Ottoman period as 'a social and political' centre and not only a commercial one (Jacobson, 2011, p. 5). Jacobson argues that the plaza in front of the gate, the municipality, and the municipal gardens, created new kinds of 'public spaces', whose symbolic importance was in representing the new shape of the Ottoman state. This was, on the one hand, a top-down development, in terms of planning as well as official use of this area for national celebrations and ceremonies. Yet as Jacobson argues, at the same time, this was a space that allowed people of various walks of life to experience Ottoman public space, to negotiate and contest the new civic order (Jacobson, 2011, pp. 54–56). Diaries and memoirs of Jerusalemites reveal that the area was used by members of all congregations and communities, and primarily men of means, who frequented the cafés, used the post offices and banks, required the services of the municipality, or strolled in the municipal gardens and listened to the Ottoman military band. They participated in patriotic demonstrations, purchased the latest fashion, and discussed politics in cafés (Wallach, 2016). The development of the new town centre was in line with similar transformations in other Ottoman cities throughout the empire that created a new urban language of modernity through familiar symbols such as flags and insignia, and landmarks such as clock towers, water fountains, and public institutions (Hanssen et al., 2002).

The town centre was also a site of social contestation. In some cities in other parts of the empire, social conflicts were often articulated in inter-communal violence, sometimes even in massacres. Civic Ottomanism did not prove sufficient to prevent such escalations, and indeed, may have contributed to them by encouraging conflicting expectations that could not be accommodated (Der Matossian, 2014; Freitag et al., 2015). However, in Jerusalem of the early twentieth century, despite emerging tensions around the question of Zionism, social contestation did not manifest in sectarian terms. Rather, conflict was articulated in terms of class differences and in tensions between local Jerusalemites and the authorities (local, district, and imperial). The Jaffa Gate plaza was an area where rich and poor mixed, known for its pickpockets and police presence.[3] One such social conflict along class and gender lines was the clampdown on peasant women selling vegetables in the city's open market in the inner plaza of Jaffa Gate. The new elite believed the unruly peasant women spoiled the image of the new town centre, and called for removal of peasant sellers from this place (Campos, 2010, p. 171). Another source of contention was plans (by the police and district council) to regulate carriage transportation, against the wishes of carriage drivers. In 1913 and 1914 there were a number of driver strikes on this issue. On one night in June 1914, carriage drivers (including Muslims and Jews) met in a local Jaffa Gate café to agitate against the new regulations. On the next day, when a driver from Bethlehem arrived at Jaffa Gate to collect passengers, the striking drivers attempted to prevent him from doing so. The matter soon descended to blows, and several drivers were arrested by the police. The local Jewish newspaper *Ha-Herut*, reported approvingly on the 'non-compromising line taken by the police to ensure regulation like in any other city'. While it reported that most Jewish drivers supported the strike, and in fact two Jewish drivers were arrested by the police for doing so, *Ha-Herut's* bourgeois sympathies, and its support for a modern and orderly town centre, clearly came before ethno-religious solidarity, at least in this case.[4]

During World War I, under the leadership of Cemal Pasha, the Ottomans escalated their attempts to control and shape the centre of town – and the larger ethos of Jerusalem. Large military processions, patriotic marches, and official ceremonies were held here to mark the Ottoman entry into the war and to celebrate the campaign against British-ruled Egypt. Allied-owned post offices and banks were closed; 'enemy languages' were banned on shop signs and adverts (Wallach, 2020). School children, of all denominations, routinely marched waving flags to show the nation's support for the military effort (Jacobson, 2011, pp. 59–60). And yet as the war efforts floundered, criticism of the Ottoman authorities was carefully discussed in local cafés (Tamari, 2011a). The Ottomans moved to more repressive means of demonstrating their rule. Jaffa Gate, the site of Ottoman modernity and civic progress, became the site of state violence and public executions. Several public hangings took place at Jaffa Gate, including that of the Mufti of Gaza, who was suspected of supporting Arab nationalist rebels, and of five soldiers who were accused of being deserters – two Christians, two Jews, and one Muslim (Jacobson, 2011, pp. 59–60). These public executions were designed to intimidate the local population, which had become increasingly disillusioned with the Ottoman state. It was a violent attempt to establish law, order, and discipline. It confirmed the significance of the town centre to the Ottoman ethos. It also manifested that both repression and resistance were not perceived in sectarian lines, of one religious group against the other, but rather in terms of broad alliances against increasingly oppressive rule.

British undoing of the Jaffa Gate centre

Unlike the Ottoman authorities and local elites, European visitors were less impressed by the new centre, as it did not fit with their ideas of what Jerusalem should look like. Pierre Loti, French Orientalist writer, visited the city in 1894 in search of 'the real Jerusalem, the Jerusalem that we have seen of old in pictures and prints'. Loti was shocked to find the town centre 'as commonplace as a Parisian suburb' (Loti, 1915, p. 172). Theodor Herzl, the Zionist leader, who visited the city in 1898, similarly disliked the mixing between old and new, sacred and profane. In his diary, he spelled out the vision for the city:

> I would clear out everything that is not sacred, set up workers' houses beyond the city, empty and tear down the filthy rat-holes, burn all the non-sacred ruins, and put the bazaars elsewhere. Then, retaining as much of the old architectural style as possible, I would build an airy, comfortable, properly-sewered, brand new city around the Holy Places.
>
> (Herzl, 1958, pp. 283–284)

The British rulers of Jerusalem, who occupied the city in December 1917, had similar sentiments. The earliest signal of the attitudes towards Jaffa Gate area could be seen General Allenby's ceremonial entry to the city, on 11 December 1917. Following detailed choreographic instructions from Mark Sykes in London, Allenby and his officers dismounted at Jaffa Gate, and entered the walled city by foot, not through the 1898 opening but through the sixteenth-century gate (Bar Yosef, 2001). This entry was designed to convey respect and reverence to the Holy City, a message that was then underlined in Allenby's proclamation at the footsteps of the Ottoman citadel, pledging to keep the religious status quo, and protect sacred places of the three religions. This message was conveyed in propaganda material as the British sought to capitalise on Jerusalem's resonance with global audiences from Ireland to India. Allenby's reverential entry on foot marked the walled city as a sacred space, to be separated and isolated from the modern secular city outside the walls. The projected sharp division between 'Old' and 'New' became a founding principle in British urban planning policy in Jerusalem (Hyman, 1994; Tamari, 2011b). This had profound implications for Jaffa Gate, which overnight transformed, yet again, from the city centre into an imaginary liminal threshold and transition point.

The British invested considerable efforts in urban planning in Jerusalem, but their efforts focused almost exclusively on the 'conservation' of the Old City. These efforts were led, in the first eight years of British rule, by Governor Ronald Storrs, who believed that the primary mission of the British was the preservation of Jerusalem as what we would call today a 'world heritage city'. This mission was far less contentious than British commitment to Zionism and the 'Jewish National Home' (which was the primary aim of the British Mandate, and met with vocal Arab opposition).

The desire to isolate the walled city from the new parts was key to British policies in Jerusalem, and was shared by virtually all British officials. They saw the late Ottoman development, especially around Jaffa Gate, as nothing but an eyesore. The Arts and Crafts writer, artist, and theorist Charles Ashbee, was recruited by Governor Storrs as a civic advisor, in charge of planning and civic improvement. Ashbee concurred entirely with Storrs's aesthetic ideals, although his political motivation was somewhat different. He hoped to use Jerusalem as his testing ground for urban ideas, in hope of saving it from the monstrous effects of industrial modernity. He was hostile to Zionism, and admired the Islamic Oriental city that he thought Jerusalem was (Ashbee, 1917; Crawford, 1985; Gitler, 2000; Hysler-Rubin, 2006). Ashbee's plans dealt with the Jaffa Gate area in great detail. His ambition was to clear the buildings outside the walls to allow for a large park system that would circle the walls and set the Old City apart. Ashbee's plans did not stop with demolishing the late Ottoman centre, which was obstructing the view of the city walls; he also recommended restoring the moat, and proposed dome-shaped caravanserais below the gate, for the Bedouins coming to the city with their camels to trade. For Ashbee, the question was not only the aesthetic appearance of the city walls, but also the preservation of a traditional way of life. But in the words of Ron Fuchs and Gilbert Hebert, by 'seeking to sustain an already-anachronistic institution, his idea was as reactionary as it was patronizing' (Fuchs and Herbert, 2001, p. 93). Ashbee mocked the modernist zeal of local Jerusalemites, and their desire to continue the late Ottoman drive towards development. '[Arab Mayor Ragheb Nashashibi] wants to make of Jerusalem a city like Paris, a continuous Champs Elysees with abundance of Kiosks,' wrote Ashbee in his *Palestine Notebooks*, 'but I tell him I am no Haussmann and we must agree to differ' (Ashbee, 1923, p. 158).

The heavy-handed British intervention in town planning was part of what Falestin Naily termed the 'de-municipalisation' of Jerusalem (Naili, 2018). The Jerusalem Municipality, which, before 1914, was the leading force in the development of the city, was assigned a decidedly minor role, as British officials seized control of planning decisions and reshaped the ethos and meaning of the city. As noted by Roberto Mazza, in the first decade of British rule, the governor, his civic advisor, the British-led Pro-Jerusalem Society, and the town planning committee left very little room for the locally elected municipality to influence decisions (Mazza, 2018). The downgrading of the municipality corresponded to the downgrading and draining of the city centre as a shared space.

The civic promise of the Ottoman modern city centre, and the fact that it encompassed and facilitated an urban non-sectarian Jerusalemite ethos, appeared to have entirely escaped the British. They saw it as ugly commercial development, an unfortunate residue of Ottoman rule. The reasons for this judgement were not only aesthetic (and the fact that the city centre hid the beautiful ancient city walls from view): it was strongly tied to the British view of Jerusalem as a deeply segregated city which could not accommodate any common urban citizenship, and therefore could not possibly have a single, shared, civic centre. The British working assumption was that the city's neighbourhoods were highly divided according to an ethnic and religious logic. They adhered to the model of the Old City as composed of four ethno-religious quarters (Tamari, 2011b).

At the same time, the civic dimensions of the Jaffa Gate centre appeared as a threat and a challenge, as a space which could host opposition and protests against British policies. The new national Arab associations, the Arab Club and the Arab Literary Club, both opened offices in the area after 1918, and they campaigned here against the Balfour Declaration (Oskotoski, 1921; al-Sakakini, 1982). In early 1920, two large Arab demonstrations and marches were held in the city centre against British commitment to Zionism. In April 1920, the Jaffa Gate plaza was the site of the first anti-Zionist riot in Palestine. The Easter or Nabi Musa riots took place during the annual Muslim pilgrimage from Hebron and Nablus to Jerusalem. The pilgrimage became the scene of political speeches and then a violent confrontation, in which Jewish shops and passers-by were attacked (Mazza, 2015). The divisive nature of British colonial policies meant that, inevitably, civic space became a space for conflict between Arabs, Jews, and colonial authorities.

The ambitious British plan to demolish much of the area and replace it with a park proved unpractical. It required the expropriation of dozens of buildings, and the compensation costs were prohibitive. As Storrs lamented in his memoirs:

> A discerning conqueror in 1850 could have established the new shops, con-
> vents and hotels well away from the Old City and have left the grey ramparts
> in a setting of grass, olives and cypresses. By 1918 the time was past for
> seeing Jerusalem adorned as a bride.
>
> (Storrs, 1945, p. 315)

But Storrs succeeded in removing the symbols of late Ottoman progress – the fountain and the clock tower. Disregarding vocal opposition from the city council, the tower, which in his words had 'too long disfigured' Jaffa Gate, was dismantled in the 1920s (Fuchs and Herbert, 2001). The removal of the clock was the most symbolic expression of the British intention to freeze time and send Jerusalem back into the past.

While the city centre buildings were not demolished, the British allowed the decline of the Jaffa Gate centre, as they drained it of civic content. The post office, the municipality, and government offices were relocated further up Jaffa Road and to the Russian compound. The British encouraged the development of Jewish commercial centres further up Jaffa Road, such as the Antymos Garden project, and the Mahne Yehuda vegetable market for Jewish shoppers. These projects diverted Jewish shoppers and businesses westward to the upper part of the street (Kroyanker, 2009, p. 46).

Escalating tensions in the city as a result of the Zionist–Arab conflict led to greater segregation. Jaffa Road developed into two sections: lower Jaffa Road was the commercial centre of Arab Jerusalem, while upper Jaffa Road was dominated by Jewish enterprises. In 1947, a Zionist guide to 'new Jerusalem' described the area between Jaffa Gate and the new municipality as an entirely 'Arab city, including its flourishing houses of commerce; because after the 1929 [riots], Jews were gradually pushed away, and they have all but disappeared from here' (Shapira, 1947, p. 111). The guide describes the municipal gardens as the transition point between the Arab 'Eastern' *al-Quds* and the Hebrew 'Western' *Yerushalayim*. Similarly, the Arab Palestinian intellectual and politician 'Arif al-'Arif, in his notes on Jerusalem shortly before the 1948 war, describes a rigid separation between Arabs and Jews. He notes that the Mamilla commercial quarter, in what was the former Ottoman centre, was entirely Arab. He also lists the city's neighbourhoods according to national and religious identity, suggesting that segregation was near total (al-'Arif, 1961, pp. 431, 469). But such clear-cut demarcation appears exaggerated, with evidence of an Arab commercial and residential presence in the upper, 'Jewish' part of Jaffa Road, as well as the Jewish presence in the 'Arab' Mamilla area. The Mamilla ('Shama'') commercial centre was developed by Sephardic and Mizrahi Jewish merchants who moved out of the Old City after 1929, but wanted to retain their Arab clientele and their ties with Arab traders (Kroyanker, 2009, p. 111).

On 2 December 1947, the Mamilla area was the site of one of the first major inter-communal clashes of the 1948 war. Arab demonstrators, protesting the UN 181 decision to partition Palestine into Jewish and Arab states, attempted to march from Jaffa Gate to Princess Mary Street and were blocked by British police. In the ensuing riots, Jewish shops were looted and burnt, Jewish militia shot into the Arab crowd; the Arab-owned Rex cinema was burnt down the following day (Collins and Lapierre, 1988; Kroyanker, 2009). In July 1948, the Israeli Defence Forces bombed many buildings in the area and effectively cut off Jaffa Gate from the west part of Jaffa Road. When the Israeli–Jordanian ceasefire came into effect, much of the Jaffa Gate area became a no man's land between the Jordanian-ruled Old City and Israeli West Jerusalem. The former city centre, the pride of the late Ottoman city, which continued to accommodate inter-ethnic interaction throughout the Mandate, had become a border zone, and remained so for 19 years (see Figure 7.2). Arab and Armenian shop owners fled to the Jordanian side of the city, leaving their property behind. Much of it stood

Figure 7.2 The border walls in Mamilla.
Source: Photo by Van de Poll, 1964. Wikimedia Commons licence: Creative Commons. CC0.

in ruin, while on the Israeli side, the Mamilla area became a slum, as somewhat-habitable buildings were used to house poor Jewish immigrant families, mostly from Arab countries. Ironically, this was the only time when the area became a residential neighbourhood. Up to 8,000 migrants lived in the Mamilla area, in former commercial buildings, in difficult conditions, and often without running water (Kroyanker, 2009). During those years of partition, the two parts of the divided city developed their respective town centres. In Israeli West Jerusalem, the Jaffa–Ben Yehuda–King George Avenue 'triangle' continued its pre-1948 development to become the centre of the city. East Jerusalem, while retaining control over the Old City, lost almost all the modern neighbourhoods, public institutions, and the main commercial area of Jaffa Road and Mamilla. It developed a new commercial centre in the area of Damascus Gate. Interestingly, the municipalities of the two parts remained in the area of the former centre. The West Jerusalem Municipality was located in the Mandatory Municipality building, while the East Jerusalem Municipality operated from a building in the inner plaza of Jaffa Gate. The two municipalities thus both appeared to make a claim to the remainder of what was once a single city (Kroyanker, 2005; al-'Asali, 2014).

The final destruction: Israeli occupation in 1967

In June 1967, Israel occupied the West Bank, including East Jerusalem. In the immediate aftermath of the war, the Israeli government decided to annex East Jerusalem unilaterally. The Jordanian East Jerusalem Municipality was disbanded. Even before the boundaries of the annexed area were decided, Defence Minister Moshe Dayan ordered the removal of all barriers between the western and eastern parts of the city. What was a frontier zone was once again about to transform, this time into a 'seam line' between the former two parts. Dayan's orders dictated a hurried timeframe, in which the border and no man's land had to be cleared by the end of June 1967.

Officials in the Israeli Jerusalem Municipality saw Dayan's clearance operation and the post-war turmoil as a golden opportunity to achieve what the British planned to do, but were never able to. They ordered the demolition of all the late Ottoman buildings adjacent to the city walls in order to separate the Old City from the new parts. Between 15 and 29 June, the municipal maintenance department raced to demolish around 200 buildings of the former Ottoman town centre between Jaffa Gate and the Mandatory Municipality building (Benvenisti, 1976, pp. 124–126). This was probably the largest demolition effort in the history of modern Jerusalem. The demolition teams worked around the clock, night and day, in order to transform the area entirely before the city was officially 'united'. By the time the former border barriers were removed, the city walls between Jaffa Gate and the north-western corner of the Old City had become visible for the first time in 90 years. Shortly afterwards, in July 1967, in a meeting between government ministers and the Jerusalem Municipality, Mayor Teddy Kolleck, and former chief of staff and archaeologist Yigael Yadin suggested declaring the areas surrounding the city walls a 'National Park' in keeping with 1920s British plans (Benziman, 1973, p. 269). Charles Ashbee's vision of the 'Park System' was finally being materialised (see Figure 7.3).

The Israeli project of Jerusalem's 're-unification' owed much to the British colonial discourse of Jerusalem. The desire to achieve visual isolation of the Old, 'Holy' City from the 'modern' town was inspired by British plans. But Israeli planners also internalised the British confessionalised view of the city as a patchwork of segregated ethno-religious communities – what Teddy Kolleck termed a 'mosaic':

> Jerusalem is no melting pot, we are not trying to make 'goulash' out of everybody, it's a mosaic of different cultures and civilisations, living in one city. This is the condition that we try to preserve and it will be Jerusalem's character also in the future. I do not think everyone will blend together and will suddenly start conversing in Esperanto […].
>
> (Kolleck, in Malchin, 2009)

While speaking in grand and lofty terms of the city's 'unification', and 'bridging East and West', in practice Israeli plans aimed to mark a clear separation between Arab-Palestinian and Jewish-Israeli areas (Pullan et al., 2007).

Figure 7.3 Old City Park, on the site of the former Ottoman post office, 2016.
Source: Wikicommons. Licence: Creative Commons. CC BY-SA 4.0

The demolition of the buildings adjacent to the wall, and the establishment of a park around the Old City, met with no resistance. This is unlike other Israeli interventions, such as the demolition of the Mughrabi quarter near the Western Wall, which caused an outcry (Institute for Palestine Studies, 1968). Israeli commentators, even those critical of official 'unification' policies, saw the park as one of Kolleck's finest achievements. One critic praised the park as a rare example in which the aggressive Judaisation agenda was sidelined by real commitment to scenic and spiritual considerations (Benziman, 1973, p. 269). David Kroyanker, a former architect in the Jerusalem Municipality and a local champion of architectural conservation, similarly justifies the demolition as a historic and necessary intervention. Kroyanker sees no civic value in this area or its conservation (Kroyanker, 2005, p. 90).

Meron Benvenisti, Kolleck's advisor and deputee played a key role in the demolition efforts as well as in the planning of the national park around the city walls. Since then, Benvenisti has become disillusioned with the Israeli annexation of East Jerusalem and with Zionism more generally, yet he continues to take pride in his part of the demolition and the creation of the park. He stresses that the planners' approach was entirely 'professional and non-political', and that efforts were made to minimise any adverse effects on the Palestinian population. Benvenisti lambasts the 'folly' and 'hubris' of his generation, whose vision was later hijacked by Jewish fundamentalists. As he points out, right-wing governments later handed over the management of the Old City national park to Jewish settler groups such as ELAD, who use the park's planning regulations in their efforts to Judaise the Palestinian area of Silwan and turn it into a biblical theme park.

With typical candidness and self-criticism, Benvenisti accepts his share of responsibility for the settler takeover of the Old City park:

> I believe that one would be justified to condemn our naivety, and worse, to argue that, in the final outcome, we are no better from the zealots of ELAD, who harass the Arabs and dispossess them in the name of "the eternal glory of Israel". We, like [them], should not have intervened in occupied territory. But should we not deserve some credit for the aesthetic value of what we created around the Old City?
>
> (Benvenisti, 2012, p. 200)

But the creation of the Old City national park was not only an aesthetic intervention, it was also a political one. The Old City walls 'park system' vision, originally put forward in the early 1920s, served several different agendas over the following century. The ideological trajectories of British planners, late 1960s Israeli officials, and early twenty-first century Israeli settlers should not be collapsed into one, and the differences between them should not be dismissed. In the 1920s, Ronald Storrs saw Jerusalem as a site of world heritage, to be protected by the British as benevolent guardians of both the Zionist revival and Arab cultural presence. Charles Ashbee sought to experiment in Jerusalem with his anti-industrial romantic ideas of the 'City on the Hill', and to 'defend' Arab Islamic culture from the invasion of electricity, automobiles, and Zionist modernism. Israeli planners arrived after 1967 with a commitment to an 'enlightened occupation': they wanted to make greater Jerusalem into a Jewish capital, while maintaining respectful and tolerant attitudes towards local Muslims and Christians (perceived as segregated religious groups rather than national indigenous populations under occupation). Whereas since the 1990s, an Israeli establishment dominated by right-wing groups has increasingly used the 'conservation' of the Old City as a mechanism to promote Jewish takeover efforts. And yet what they all share is the colonial Orientalist view that dismissed the civic vision embodied in the Ottoman city centre, based on local non-sectarian governance. All ignored the wishes of the native population of Jerusalem in favour of an imperial vision; they sought to reverse the mundane modernisation and impose the weight of history onto the city. They also assumed that shared civic space and a common sense of Jerusalemness is impossible in a city that is by its nature segregated and divided. The success of British and Israeli officials was not only in forcing upon the city its "original" meaning, by reshaping the built environment, but was also in the effacement of the Ottoman civic town centre from the historiography and cultural memory of Jerusalem. The inscription of confessionalised space into Jerusalem's landscape, in a manner that naturalised it as historical legacy and inevitable present and future, required the removal of urban citizenship, both physically and metaphorically.

After the Jaffa Road buildings near the wall had been destroyed, Israeli officials prepared the ground for a flagship revival project of the nearby Mamilla Street. The Mamilla regeneration was trumpeted as a vehicle to connect West and East Jerusalem and to cement the unification of the city (Kroyanker, 2009; Nitzan-Shiftan, 2017). The project was completed only in the 2000s, at considerably reduced dimensions from the original plans. It is beyond the scope of this chapter to review in detail the transformation of this project. It is interesting to note, however, that the Mamilla shopping arcade bears some resemblance to the Ottoman town centre: with hotels, cafés, and luxury brands, it is one of the most globalised malls in Jerusalem today, and in many ways, the one that is most mixed in national terms (Shtern, 2016). Because of its proximity to the Old City and its central location on the North–South route (Road No. 1), it is highly accessible to Arab Palestinians who make up a significant presence as shoppers, workers (shop assistants, waiters, kitchen staff, and cleaners), as well as pedestrians using the arcade as a route into the Old City. However, such commercial 'shared space' has no civic undertones; the project was designed and is owned by Jewish Israelis. The Mamilla arcade was designed in stylised Orientalist style, but makes no reference to the Ottoman town centre, or to Arab heritage, such as the ancient Mamilla Cemetery nearby. The handful of buildings chosen for conservation were the ones associated with the Zionist and European histories of the area, and not buildings with Arab or even local Jewish history (Kroyanker, 2009, p. 262). The dramatic modern history of this area is not mentioned anywhere in the arcade or near Jaffa Gate: the celebrations of Ottoman constitutional revolution in 1908; the hanging of dissidents and army deserters in 1916; the anti-Zionist riots of 1920; and the Arab–Jewish clashes of December 1947. All of these have been erased, along with the buildings and spaces where they took place. The undeniable success of the destruction of the late Ottoman centre of Jerusalem is expressed in its parallel erasure from the literature on Jerusalem. British and Israeli planners thus succeeded in the perfect crime: not only were they able to demolish the Ottoman urban fabric and the civic vision it embedded, but they were also able to remove this vision from public memory.

Arliella Azoulay argues that house demolition is a primary trait of the Israeli regime, a strategy in the service of the Judaisation of urban space (Azoulay, 2013). These policies are characterised by takeover of space and the exclusion of Palestinians by forced removal, military rule, or social exclusion. Thus, Israeli-designed 'public space' – such as the Old City park and the Mamilla arcade – appears flawed as it fails in its 'most important principle' – to be open to all for participation. In Azoulay's view, these policies can be challenged by the development of 'civil imagination' which hinges on 'a being together of individuals, and not as a product of the governing power' – that is, the ability to imagine Jews and Arabs together, as a single civic group. The Ottoman town centre provides one of the more dramatic examples of colonial urban destruction of public space, envisaged by the British and implemented under Israeli rule. But it also points to the limits of Azoulay's concept of a non-hegemonic civil imagination of 'being together', which could circumvent the Arab/Jewish dichotomy. The Ottoman public space of Jaffa Gate was, in fact, a 'product of the governing power'. The Ottoman city centre was developed by a political regime that allowed for a shared identity, non-sectarian citizenship, and a local multi-faith elite. This is not to say that public space was merely a top-down creation, but rather that the Ottoman framework allowed for both power and resistance to operate on the basis of local non-sectarian identity. After 1917, the absence of such a common framework made civil imagination virtually impossible. Perhaps the best example of this is the erasure of the town centre from both Israeli and Palestinian cultural memory of modern Jerusalem. This is precisely because the centre was Ottoman, Jerusalemite civic space which was neither 'Jewish' nor 'Arab', and therefore does not serve the dominant nationalist narratives. The possibility of imagining a common past, and a Jerusalem not defined by segregation, requires expanding the frame beyond the nationalist narratives of the conflict.

Notes

1 Since 2000, affluent Palestinian families have moved into 'Jewish' neighbourhoods, but this remains a limited phenomenon (Yacobi and Pullan, 2014).
2 Kremer, Mendel. 'Ha-Shavua', *Hashkafa,* 29 July 1904.
3 'Yerushalayim Yom Yom', *Ha-Tsvi,* 3 September 1912.
4 'Yerushalayim', *Ha-Herut,* 19 June 1914.

References

Historical press

[Author unknown]. 19 June 1914. 'Yerushalayim.' *Ha-Herut*.
[Author unknown]. 3 September 1912. 'Yerushalayim Yom Yom.' *Ha-Tsvi*.
Kremer, Mendel. 29 July 1904. 'Ha-Shavua.' *Hashkafa*.

Literature

al-'Arif, 'Arif. 1961. *al-Mufassal fi tarikh al-Quds*. Jerusalem: Matba'at al-Ma'arif.
al-'Asali, Kamil. 2014. '*al-Quds fi al-Fitra al-Urduniyya*.' *Akhbar el-Balad*, Accessed 7 September 2019. http://akhbarelbalad.net/ar/1/2/1330/?fbclid=IwAR2z40PjXvzleV112E43cl5 Tqi22tdLI_BkEFzVzd2o9kASNa6HRy7eiFKc.
al-Sakakini, Khalil. 1982. *Kadha ana ya dunya*, 2nd ed. Beirut: al-Ittihad al-'amm lil-kuttab wa-al-suhufiyin al-filastiniyin.
Ashbee, Charles Robert. 1917. *Where the Great City Stands: A Study in the New Civics*. London: The Essex House Press.
Ashbee, Charles Robert. 1923. *A Palestine Notebook: 1918–1923*. London: William Heinemann.
Auld, Sylvia, and Robert Hillenbrand, eds. 2000. *Ottoman Jerusalem: The Living City, 1517–1917*. London: Altajir World of Islam Trust.

Azoulay, Ariella. 2013. 'When a Demolished House Becomes a Public Square.' In *Imperial Debris: On Ruins and Ruination*, edited by Ann Laura Stoler, 194–224. Durham, London: Duke University Press.

Bar Yosef, Eitan. 2001. 'The Last Crusade? British Propaganda and the Palestine Campaign, 1917–1918.' *Journal of Contemporary History* 36 (1): 87–109.

Ben-Arieh, Yehoshua. 1984. *Jerusalem in the 19th Century: The Old City*. New York: St. Martin's Press.

Ben-Arieh, Yehoshua. 1986. *Jerusalem in the 19th Century: Emergence of the New City*. Volume 2. New York: St. Martin's Press.

Gitler, Inbal Ben-Asher. 2000. 'C.R. Ashbee's Jerusalem Years: Arts & Crafts, Orientalism and British Regionalism.' *Assaph: Studies in Art History* 5: 29–52.

Benvenisti, Meron. 1976. *Jerusalem: The Torn City*. Jerusalem: Isratypeset.

Benvenisti, Meron. 2012. *Halom ha-tsabar ha-lavan: otobiyografyah shel hitpakhhut*. Jerusalem: Keter.

Benziman, 'Uzi. 1973. *Yerushalayim: 'ir le-lo homah*. Jerusalem: Schocken.

Büssow, Johann. 2011. *Hamidian Palestine: Politics and Society in the District of Jerusalem 1872–1908*. Leiden, NL: Brill.

Calame, Jon, Esther Charlesworth, and Lebbeus Woods. 2009. *Divided Cities: Belfast, Beirut, Jerusalem, Mostar, and Nicosia*. Pittsburgh: University of Pennsylvania Press.

Campos, Michelle. 2010. *Ottoman Brothers: Muslims, Christians, and Jews in Early Twentieth-Century Palestine*. Stanford: Stanford University Press.

Chapman, Rupert L. 2018. *Tourists, Travellers and Hotels in 19th-Century Jerusalem: On Mark Twain and Charles Warren at the Mediterranean Hotel*. London and New York: Routledge.

Chelouche, Yosef Eliyahu. 2005. *Parashat hayay 1870–1930*. Tel Aviv: Babel.

Collins, Larry, and Dominique Lapierre. 1988. *O Jerusalem*. New York: Simon and Schuster.

Crawford, Alan. 1985. *C. R. Ashbee: Architect, Designer and Romantic Socialist*. New Haven: Yale University Press.

Der Matossian, Bedross. 2014. *Shattered Dreams of Revolution: From Liberty to Violence in the Late Ottoman Empire*. Stanford: Stanford University Press.

Dimitriadis, Sotirios. 2018. 'The Tramway Concession of Jerusalem, 1908–1914: Elite Citizenship, Urban Infrastructure, and the Abortive Modernization of a Late Ottoman City.' In *Ordinary Jerusalem 1840–1940*, edited by Angelos Dalachanis and Vincent Lemire, 475–489. Leiden, NL: Brill.

Freitag, Ulrike, Nelida Fuccaro, Nora Lafi, and Claudia Ghrawi, eds. 2015. *Urban Violence in the Middle East: Changing Cityscapes in the Transformation from Empire to Nation State*. New York: Berghahn Books.

Freitag, Ulrike, and Nora Lafi. 2014. *Urban Governance Under the Ottomans: Between Cosmopolitanism and Conflict*. London and New York: Routledge.

Fuchs, Ron, and Gilbert Herbert. 2001. 'A Colonial Portrait of Jerusalem, British Architecture in Jerusalem of the Mandate 1917–1948.' In *Hybrid Urbanism: On the Identity Discourse and the Built Environment*, edited by Nezar AlSayyad, 81–108. Westport, CT: Praeger.

Gilbert, Martin. 1996. *Jerusalem in the Twentieth Century*. London: Chatto & Windus.

Glass, Joseph B., and Ruth Kark. 2018. 'The Jerusalem Chamber of Commerce, Industry, and Agriculture, 1909–1910: An Early Attempt at Inter-communal Cooperation.' *British Journal of Middle Eastern Studies* 45 (2): 269–289.

Hanssen, Jens, Thomas Philipp, and Stefan Weber. 2002. *The Empire in the City: Arab Provincial Capitals in the Late-Ottoman Empire.* Beirut: Wurzburg Ergon Verlag.

Herzl, Theodor. 1958. *The Diaries of Theodor Herzl,* edited by Marvin Lowenthal. London: Victor Gollancz.

Hyman, Benjamin. 1994. 'British Planners in Palestine, 1918–1936.' PhD thesis, London School of Economics, Department of Geography.

Hysler-Rubin, Noah. 2006. 'Arts & Crafts and the Great City: Charles Robert Ashbee in Jerusalem.' *Planning Perspectives* 21 (4): 347–368.

Institute for Palestine Studies. 1968. *The Rights and Claims of Moslems and Jews in Connection with the Wailing Wall at Jerusalem.* Beirut: The Institute for Palestine Studies.

Jacobson, Abigail. 2011. *From Empire to Empire: Jerusalem Between Ottoman and British Rule.* Syracuse, NY: Syracuse University Press.

Jawhariyyeh, Wasif. 2013. *The Storyteller of Jerusalem: The Life and Times of Wasif Jawhariyyeh, 1904–1948.* edited by Salim Tamari and Issam Nassar. Northampton, MA: Interlink Publishing.

Kark, Ruth, and Michal Oren-Nordheim. 2001. *Jerusalem and Its Environs: Quarters, Neighborhoods, Villages, 1800–1948.* Jerusalem: Hebrew University Magnes Press.

Kroyanker, David. 2005. *Reḥov Yafo, Yerushalayim: biyografyah shel reḥov: sipurah shel 'ir.* Yerushalayim: Mekhon Yerushalayim le-ḥeḳer Yiśra'el: Keter.

Kroyanker, David. 2009. *Mamila: ge'ut, shefel ye-hitḥadshut i rova' Alrov-Mamila.* Jerusalem: Keter.

Lemire, Vincent. 2017. *Jerusalem 1900: The Holy City in the Age of Possibilities.* Chicago: University of Chicago Press.

Loti, Pierre. 1915. *Jerusalem.* London: T. Werner Laurie.

Malchin, Ya'acov. 2009. 'Interview with Teddy Kollek, 1984 [Hebrew].' *Yahadut Hofshit* 30.

Mazza, Roberto. 2015. 'Transforming the Holy City from Communal Clashes to Urban Violence, the Nebi Musa Riots in 1920.' In *Urban Violence in the Middle East: Changing Cityscapes in the Transformation from Empire to Nation State,* edited by Ulrike Freitag, Nelida Fuccaro, Nora Lafi and Claudia Ghrawi, 179–196. New York: Berghahn Books.

Mazza, Roberto. 2018. '"The Preservation and Safeguarding of the Amenities of the Holy City without Favour or Prejudice to Race or Creed": The Pro-Jerusalem Society and Ronald Storrs, 1917–1926.' In *Ordinary Jerusalem, 1840–1940,* edited by Angelos Dalachanis and Vincent Lemire, 403–422. Leiden, NL: Brill.

Montefiore, Simon Sebag. 2012. *Jerusalem: The Biography.* New York: Penguin.

Naili, Falestin. 2018. 'The De-Municipalization of Urban Governance: Post-Ottoman Political Space in Jerusalem'. *Jerusalem Quarterly* 76: 8–13.

Nassar, Issam. 2003. "Early Local Photography in Jerusalem.' *History of Photography* 27 (4): 320–332. doi: 10.1080/03087298.2003.10441267.

Nitzan-Shiftan, Alona. 2017. *Seizing Jerusalem: The Architectures of Unilateral Unification.* Minneapolis, MN: University of Minnesota Press. Accessed 31 July 2019. www.jstor.org/stable/10.5749/j.ctt1pv88z4.

Oskotoski, Dfus levi. 1921. *Kol yerushalayim le-shnat tarpa-b 1921: Sefer shimushi, sefer yedi'ot, sefer adrese'ot.* Accessed 4 February 2019. Jerusalem: Dfus Levi.

Pullan, Wendy, Philipp Misselwitz, Rami Nasrallah, and Haim Yacobi. 2007. 'Jerusalem's Road 1: An Inner City Frontier?' *City* 11 (2): 176–198.

Schölch, Alexander. 2002. 'Jerusalem in the 19th Century 1831–1917 AD.' In *Jerusalem in History: 3000 BC to Present,* edited by Kamil Jamil Asali, 228–248. London: Kegan Paul.

Shapira, Yitzhak. 1947. *Yerushalayim Mihutz Lahoma: Tolada vehavai*. Jerusalem: Palestine Book Press.

Sheehi, Stephen. 2015. 'Portrait Paths: Studio Photography in Ottoman Palestine.' *Jerusalem Quarterly* 61: 23.

Shtern, Marik. 2016. 'Urban Neoliberalism vs. Ethno-national Division: The Case of West Jerusalem's Shopping Malls.' *Cities* 52: 132–139.

Storrs, Ronald. 1945. *Orientations: Ronald Storrs*, Definitive 2nd ed. London: Nicholson & Watson.

Tamari, Salim. 2009. *Mountain Against the Sea: Essays on Palestinian Society and Culture*. Berkeley, CA: University of California Press.

Tamari, Salim. 2011a. *Year of the Locust: A Soldier's Diary and the Erasure of Palestine's Ottoman Past*. Berkeley, CA: University of California Press.

Tamari, Salim. 2011b. 'Confessionalism and Public Space in Ottoman and Colonial Jerusalem.' In *Cities and Sovereignty: Identity Politics in Urban Spaces*, edited by Diane Emily Davis and Nora Libertun de Duren, 59–82. Bloomington: Indiana University Press.

Tamari, Salim. 2017. *The Great War and the Remaking of Palestine*. Berkeley, CA: University of California Press.

Tsoar, Dana, and Ran Aaronsohn. 2006. 'Miderekh Historit lerehov Rashi: Itsuvo upituho shel Rehov Yafo biyerushalayim 1860–1948.' *Kathedra* 121: 101–136.

Wallach, Yair. 2016. 'Jerusalem between Segregation and Integration: Reading Urban Space through the Eyes of Justice Gad Frumkin.' In *Modernity, Minority, and the Public Sphere: Jews and Christians in the Middle East*, edited by S.R. Goldstein-Sabbah and H.L. Murre-van den Berg, 205–233. Leiden, NL: Brill.

Wallach, Yair. Forthcoming 2020. *A City in Fragments: Urban Texts in Modern Jerusalem*. Stanford, CA: Stanford University Press.

Wishnitzer, Avner. 2015. *Reading Clocks, Alla Turca: Time and Society in the Late Ottoman Empire*. Chicago: University of Chicago Press.

Yacobi, Haim, and Wendy Pullan. 2014. 'The Geopolitics of Neighbourhood: Jerusalem's Colonial Space Revisited.' *Geopolitics* 19 (3): 514–539.

Yehoshua, Jacob. 1981. *Yerushalayim Temol Shilshom: Pirkey Havai*. Volume 3. Jerusalem: Rubin Mass.

Part III

Intimacies

Neighbourhoods as sources
and objects of claim-making

8 Violence, temporality, and sociality

The case of a Kashmiri neighbourhood

Aatina Nasir Malik

Introduction

It was 13 August 2017, and despite being a Sunday, the market and streets in Nowhatta were heavily crowded. People were purchasing vegetables, fruit, mutton, curd, and other groceries as if the shops were never going to open again. The roads were hectic with people trying to hastily bargain and make purchases before the stocks were exhausted. Pakistan and India were to celebrate their Independence Days on the two following days and Kashmir, which Pakistan calls its 'jugular vein' and India its 'integral part', was to suffer two days of restrictions in some areas and two days of strict curfew in other areas including parts of downtown Srinagar. On the same day, the Hizbul Mujahidin commander Yasin Itoo (popularly called Ghaznavi) was killed along with two other militants – Sheikh Irfan and Umar Majid – in south Kashmir's Shopian district in a gunfight with the Indian armed forces. On 14 August, Nowhatta, being the centre of resistance activities that include episodes of stone-throwing (*kanni-jung*) against armed troops, effectively became a prison. The entry/exit points on all sides were blocked via barricades and concertina wires and the armed troops allowed no movement. The curfewed Nowhatta that had boomed with life the day before was blanketed in silence on the outside, while inside it brewed with anger and pain. A video in which Ghaznavi's mother is speaking of her son's martyrdom and thanking God for making him the chosen one quickly went viral in the neighbourhood, with most people talking about it, asking each other if they had seen it, and passing around their mobile phones to let everyone have a glimpse. The video led to a wide range of emotions: some cried and were not even able to finish watching the video; some prayed for freedom (*azadi*); others became angry; some fell into a long silence; and some chose not to watch it at all, describing such deaths as an everyday occurrence. The escalated emotions forged a collective, although outdoor assemblies were banned. It became clear that the state could enforce a curfew on the streets but could not control the circulation of sentiments.

On 15 August (India's Independence Day), the mobile internet and communication facilities were cut off as well, so people mostly turned to television to keep informed about state violence (*halaat*). By late afternoon, however, there was a commotion of a different sort as food supplies in some of the households were running low and women became concerned about that night's meal. They began trading what they had, while others had to make do with just rice and lentils that night. As the Indian troops began to retreat to their camps in the early evening, this was followed by some stone-throwing. However, even after the curfew was relaxed, no one came out onto the streets or to the markets, thus forging a collective rejection of the normality of India's presence in Kashmir.

The ethnographic vignette above invites an exploration of the everyday of a neighbourhood to establish the ways in which violence and ordinary life interact, are separated, or blend into each other. The neighbourhood in discussion, Nowhatta, appears as a monolith in the dominant media and literary representations. The Indian narratives describe it as a hub of stone-throwers who are the paid agents of Pakistan fighting its proxy war in India. The local narratives contend that they are the victims and resisters of state violence. The chapter departs from both narratives that, in their own ways, over-determine the spectacular violence to instead explicate the everyday lived realities by understanding the relationship between violence, temporality, and sociality. It focuses on solidarity, negotiation, and discord amidst the shifting boundaries of collective political subjectivity and personal subjectivity vis-à-vis the state of occupation. Borrowing from the work of Adam Roberts, Duschinski et al. (2018) argue that the process of occupation is context specific and cannot be reduced to a singular 'character and purpose', thus being multifaceted in Kashmir as well. This chapter looks at occupation and state violence through the imposition of curfews, which not only prevent mobilization for pro-freedom politics but also entail a collective punishment of Kashmiris by hampering their everyday routines. With the manifestation of occupation and state violence in various forms, the chapter attempts to challenge the assumption that everyone in the neighbourhood is impacted by violence in the same way, implying that everyone resists, is resilient, or reacts in a similar way, thus defying the understanding of the neighbourhood as a monolith.

The chapter is based on ethnographic fieldwork[1] conducted in Nowhatta from May 2017 to December 2017 and on oral testimonies of the inhabitants to understand the relationship of the porous and non-porous boundaries between violence and the everyday, and between ordinary and extraordinary experiences. The fieldwork involved participant observation in homes, on the streets, lanes, and at markets. I would sometimes accompany my female interlocutors to work, hospital, prayers, and shopping. The interviews were conducted and testimonies collected in interlocutors' homes where they would not only unfurl their narratives of living with and negotiating violence but also expose the ruptures within community experiences.

As far as neighbourhood practices and relations are concerned, anthropology has explored neighbourhoods following event(s) of violence or natural disaster(s) showing both how resettlement is a social rather than state-directed process (Ibrahim 2018) and the importance of social capital in rebuilding neighbourhoods (Chamlee-Wright and Storr 2011). In the context of political violence, there has been work that looks at inter-ethnic relations within neighbourhoods before and after event(s) of violence, especially with a focus on the destruction and re-construction of social ties and neighbourhood relations

(Das 2007; Jasani 2008; Robinson 2013; Lofranco 2017). However, beyond what Brubaker (2002) calls *ethnic code biasing*, which tends to privilege one identity (communal identity) over others, thus obscuring the complexity of everyday social processes and relations, scholars have also looked at neighbourhoods from other perspectives. These include Sorabji's (2008) work on a Bosnian neighbourhood where neighbourhood practices and feelings are considered a 'duty' in the Islamic tradition, or Ring's (2006) work in Karachi apartments that elucidates peace as not simply the absence of violence but a cultural process that one has to constantly work towards.

The chapter deals with everyday neighbourhood practices and relationships in Nowhatta when violence is embedded in the very fabric of life. In Kashmir the violence is ongoing and is manifested through killings, injuries, baton charges, arrests, torture, cordon and search operations, humiliation, curfews, surveillance, etc. It is thus impossible to speak about life before and after the violence; rather, with the continuity of violence, it is difficult to separate parts of life that are untouched by it. In such a situation, how does one talk about the spectacular and the everyday, the ordinary and the extraordinary in a Kashmiri neighbourhood?

The first part of the chapter locates Nowhatta in the history of the Resistance Movement in Kashmir underlining how the neighbourhood spatially and materially paves the way for resistance through various means, challenging the dominant state and local narratives that focus mostly on 'stone-throwing'. Then, based on oral testimonies, it discusses the *curfew of 2016* in Nowhatta to understand the links between violence, temporality, and sociality. It unravels how, during the times of curfew, networks of solidarity and new spaces of sociality are created, highlighting the power of the local in sustaining the ordinary, while offering a lens to explicate the relationship between ordinary and extraordinary experiences and times. And finally, there is an exploration of the disruptions and fractures within the community based on personal subjectivities, and of how these personal sufferings are endured for the sake of collective political community.

Locating Nowhatta in the history of resistance in Kashmir

The Indian occupation in Kashmir began soon after the decolonization of the Indian subcontinent by the British in 1947, leading to the formation of two separate nation states, i.e. India and Pakistan. The fate of Kashmir, then an independent state with a Muslim majority, was left undecided and a plebiscite was to be held to allow the people of Kashmir to choose their side. However, in October 1947 with the coming of Pakistan-backed raiders to aid the indigenous revolt against the Hindu Dogra King, the Dogra rulers of Kashmir, without considering Kashmir's political aspirations for self-rule, signed the Instrument of Accession with India, which was marked by the sending of Indian armed troops to Kashmir. It initiated a military conflict with Pakistan, leading to four wars over the years. The plebiscite that was promised at the time of accession by both India and the United Nations Security Council became increasingly inconvenient for India and remains pending. It is something that Kashmiris continue to demand.

The situation worsened when the 1987 elections were rigged, preventing a popular political party that raised issues such as self-determination in the political arena from acquiring legislative power. The members of this party felt cheated and took to armed rebellion with support in the form of training and ammunition from Pakistan in the 1990s to gain freedom (*azadi*) from Indian rule (see Schofield 2000; Bose 2011). The region soon became the 'most heavily militarised place in the world', with close to 750,000 Indian soldiers controlling a population of roughly 9 million Kashmiris (Kazi 2010, 97). As Junaid argues, a totalizing counterinsurgency grid, underpinned by a logic of 'security' and 'control', was laid over the region which not only prevented protests but also created elaborate restrictions on everyday life, pushing Kashmiris further toward armed struggle (Junaid 2019, 5). The indiscriminate use of bullets, pellets, tear gas shells, arrests, torture, etc. at the same time unleashed the gravest human rights violation on the people of Kashmir. Duschinski (2009) argues that Kashmir has been rendered a site of confinement by these strategies – an everyday prison for the people of Kashmir. The collective memory of the denial of Kashmir's desire for self-determination continues to shape Kashmiri political consciousness and identity (Duschinski et al. 2018).

Nowhatta is located in the downtown area of Srinagar, some 5 km from the city centre, Lal Chowk, and houses various places of significant importance including the historic Jamia Masjid or the Grand Mosque which is the biggest mosque in Kashmir. Downtown, also called *sher-e-khaas*, comprises the oldest areas of the city and is central to the pro-freedom or resistance movement known locally as *Tahreek*. Nowhatta is a part of the downtown area and consists of several sub-*mohallas*, named for the usual occupation of the inhabitants of that sub-*mohalla* in the past. These sub-*mohallas* are located to the west (for example, Jabgaripora, Ranghaman, Razgaripora, Sheikh Colony) and east (for example, Roshangar Mohalla, Pyal Gar Mohalla, Malaratta) of *Syed pora*. *Syed pora* is the central street that has most of the retail shops, a police station, a park, a bank, and runs into the main *chowk* (a junction where four roads intersect and which is marked by a fenced fountain), locally named *Shaheed Muntazir Chowk* after Muntazir, a stone-thrower who was killed during stone-throwing by the government forces in 2007. The Nowhatta *chowk* is recognized as the main site of stone-throwing and has been the seat of resistance for decades.

In Nowhatta, every Friday following the afternoon prayers there is a routinized stone-throwing in which young boys and men attack the armed government forces with stones, who in turn fire bullets, pellets, tear gas shells, and launch baton charges (*lathi* charges) at the protestors. The stone-throwing in Kashmir is largely believed to be influenced by the Palestinian Intifada, in which Palestinian protestors throw stones at the Israeli armed forces. However, the older generation contends that stone-throwing is not an innovation of the 1990s, but that the members of the local rival political factions (called *shers* and *baqras*) threw stones at each other in the early 1930s. However, since 1989, the *chowk* has been the locus of resistance against the Indian state as stone-throwing begins there and then travels into the smaller streets of the neighbourhood. Young masked and unmasked men attack the troops stationed on the streets adjoining the *chowk* and, depending upon the intensity in terms of number and fervour, they go on to pelt stone at the police station at the other end of the *chowk*. Some of these men carry flags with slogans such as 'We want freedom' (*hum kya chahte azadi*), 'There's no God but Allah' (*la-ila-ha il-lal-lah*), and 'Go India go back'. The troops fire bullets, pellets, and mostly tear gas shells, affecting the neighbourhood's population. These exchanges grip the neighbourhood for 1–2 hours, with impacts lingering in various forms for a much longer period. The neighbourhood has become 'famous' as a stone-throwing neighbourhood, with both local and national media playing an active role in disseminating images of stone-throwing (mostly masked) youngsters. The Indian media, especially the prime-time news shows, talk about them as unemployed, wayward youth who are a fringe group or paid agents of Pakistan. Local print and online media instead publicize the injuries inflicted on the stone-throwers, and in other cases present those involved as rebellious, defiant stone-throwers standing up to the oppressor.

In the 2000s, even if there was a lull in the pro-freedom movement in terms of the activities of the armed militants, stone-throwing continued every week after Friday afternoon prayers. On 8 July 2016, the Hizbul Mujahidin commander Burhan Wani, who was popular among Kashmiris because of his social media presence, was killed in a gunfight with the Indian armed forces in the Kokernag area of south Kashmir along with two of his comrades – Sartaj Ahmed and Parveiz Ahmed. Following that, the entire valley was overwhelmed with the pro-freedom sentiment with protests being met by bullets and pellets from the Indian troops, resulting in many protesters being injured or killed. There was also a marked increase in the number of local boys joining the armed militancy and being killed in gunfights with the government troops. Stone-throwing as an act of resistance became more frequent in rural Kashmir, where civilians would throw stones at troops to rescue the trapped militants. Nowhatta, although quite distant from the sites of gunfights most of the time, continued witnessing stone-throwing incidents.

Neighbourhood – a space of dissent

This section illustrates some of the ways in which resistance other than stone-throwing marks the neighbourhood, explicating the materiality and spatiality of resistance. It explores how resistance is felt in physical space through its enactment and its remains, posing a challenge to the narratives of the state and thereby shaping collective memory, community, and a collective identity.

Nowhatta not only responds to excesses of the state in its vicinity but also to violence that occurs throughout the valley. The situation in Nowhatta in terms of pro-freedom activities, then becomes the locus to gauge state violence (*halaat*) in the entire valley. That perhaps is the reason that Nowhatta and its adjoining areas are often under a curfew or restrictions imposed by the state. This not only obscures the state violence behind the facade of maintaining 'law and order' but also leads to propagation of the image of a 'normal touristy' Kashmir outside.

Curfew is one of the most significant state responses to pro-freedom politics that hinders movement of people outside their homes, preventing protest, and is usually accompanied by e-curfews, i.e. blocking of internet service. Junaid (2019) argues that

> curfews are the state's default response to Kashmiri counterclaims over public space and are based on an inherent suspicion that every native assembly is a potential threat. By concentrating armed soldiers in sheer numbers within a vast militarized infrastructure, the state produces an affective geography of fear, which becomes fully materialized during curfews.
>
> (Junaid 2019, 6)

A curfew could be imposed in the entire Kashmir valley or specifically in the area where the violent event has occurred, or in areas active in resistance activities via protests and stone-throwing. Curfew is imposed in different situations: for example, when a popular militant is killed, on the death anniversary of a popular militant, or on the anniversary of a massacre/mass killing by state forces in Kashmir. Curfew is sometimes accompanied by certain relaxations allowing movement via curfew passes in case of emergencies – especially a medical emergency. However, as Anjum and Varma (2010) write, there have been instances when such passes fail to be accepted, leading to further violence at the hands of the troops.

In Kashmir, there are both expected curfews (e.g. on the death anniversary of a popular militant or other days of political importance) and unexpected ones which end up being of a longer duration, like the curfews of 1990, 2008, 2010, and 2016. Unexpected curfews are usually prompted by the killing of a popular militant or a pro-freedom leader, or sometimes a civilian by the Indian armed forces. Such killings lead to massive protests and stone-throwing, and curfew therefore becomes the state's tool to keep resistance under control. Nonetheless, curfews are frequently defied by protestors and stone-throwers, leading to the perpetuity of state violence not only through bullets, pellets, tear gas shelling, and detentions but also through sustenance of the curfew for longer periods.

Curfew is imposed by the state, but one of the important means of resistance is the observance of *hartal*, a call to strike issued by the pro-freedom group(s), and which has a very similar spatial and material effect. Shops are closed, streets become emptier. While a curfew is enforced by the military infrastructure of the state by barricading the roads or having troops patrolling the streets to curb movement, *hartal* as a tool of protest entails a kind of self-regulation of people's movement outside their homes in protest of state excesses. The power of the *hartal*, unlike curfew, comes from a collective belief in pro-freedom politics making it an act of collective resistance against collective suffering with no visible external force imposing it. Following the killing of Burhan Wani in 2016, with more and more youngsters joining the armed militancy and being killed, and with escalating state violence in different forms, the observance of *hartal* has been so consistent and normalized that people have now started to distinguish them by degree, as people are often heard asking: 'Is today an unrestricted or a restricted *hartal*?' (*az cha hartal yalai ya band?*).

Besides *hartal*, martyrs' taps (*shaheed nalq-e*) constructed in the memory of local neighbourhood militants killed by Indian armed forces in different parts of Kashmir can be found in the smaller streets (see Figure 8.1). These memorials consist of a tap with running water, a tiled or cemented sink around it, and an epitaph with the name, father's name, date of birth, affiliation (with the militant

Figure 8.1 Martyr's tap (*Shaheed Nalq-e*) in the memory of a neighbourhood militant.
Source: Photo by Aatina N. Malik, October 2017.

outfit), and place of death of the militant. There may also be a verse from the Quran hailing the militant for his sacrifice or bravery. The construction of these taps is via a collective fund from the neighbourhood as a sign of resistance in the public domain.

Graffiti in the form of slogans for freedom is quite visible in public spaces in the neighbourhood, i.e. on shop shutters, walls of schools, shrines, etc. The slogans read, '*Kashmir ki azadi tak jung rahegi*' (The war shall continue until Kashmir's freedom), 'We want freedom', 'Go India go back', etc. These slogans are hastily written without any attention to aesthetics, but have a deeper meaning and significance. In the context of Palestine, Peteet (1996) talks about graffiti art as a cultural production that not only inverts the dominant–subordinate position but also attacks the surveillance state, which cannot reach every place. In Kashmir, the graffiti and taps are acts and objects of resistance that manage to circulate information and knowledge against the will of the state. Urdu slogans are written in Latin script – for example, '*Hum Kya Chahte, Azadi!*' (We want freedom). This, borrowing from Said (1983), exemplifies the conception that no one writes simply for oneself, there is always the 'other'. These slogans are written for the 'other' who otherwise denies you a voice. However, the way this 'other' receives the writings on the walls presents an interesting dynamic in the case of Kashmir. The slogans – for example, those written on the walls of the shrine of Bah-ud-din Sahib – were originally written in black ink. However, the black ink has been overwritten with blue ink, obscuring the slogans completely or changing the letters here and there so that the meaning of the slogan is changed (see Figure 8.2).

Resistance in Kashmir comes at the price of being arrested, tortured, or interrogated. These acts of resistance thus become meaningful because they not only disseminate a message of freedom against the state's wishes, but also reduce the chance of one or two individuals being held responsible for such anonymous resistance acts, somewhat reducing the danger of arrest or torture. In such cases where it's difficult to make arrests, the state instead destroys the 'creations of resistance', attempting to render them meaningless; nevertheless, the meaning remains.

The bakers' and other storefronts (*wan-e pind*) in the neighbourhood are important for understanding the ways in which the neighbourhood gives space to resistance. Political talks and discussions take place early every morning at the bakers' shops. Breakfast in Kashmir normally includes bread bought from the bakers, so with the morning bread the men also get news and a chance to take part in political discussions, making the bakers a spot for stories, news, predictions, and rumours. It is said that the bakers' shops are *the BBC of the neighbourhood* that keeps the neighbourhood informed. Groups of men also chat at other storefronts, usually after the curfew when the armed troops have retreated to their camps. My position as a female researcher only let me observe these collectives of men every now and then; it was through interviews that I later gathered that these men talk about a wide range of topics ranging from the government, to the army, stone-throwers, militants, news, political trends on Facebook, etc. marking them as alternative spaces of political opinion and resistance.

Figure 8.2 Graffiti on the wall of the shrine of Bah-ud-din Sahib.
Source: Photo by Aatina N. Malik, October 2017.

Solidarities and new spaces of sociality

The neighbourhood as a lived space acquires significant meaning during a curfew. During my fieldwork in the neighbourhood, the curfews were not too long. They would last for two days at most, bringing along their own share of difficulties, such as the sick not being able to reach the hospital, people not being able to go for condolence meetings, and the feeling of being trapped inside their houses. However, the difficulty one faces during a curfew and the negotiations surrounding them were mostly discussed in reference to the curfew of 2016. In 2016, as already mentioned, the Hizbul Mujahidin commander Burhan Wani, along with two of his comrades, was killed in a gunfight on 8 July. Following his death, widespread protests kept the uprising going for nearly half a year, in which approximately 100 civilians died and over 15,000 civilians were injured. It was one of the worst uprisings, with Kashmir placed under almost two months of curfew that was lifted only in the first week of September 2016.

In order to understand the relationship between the ordinary and extraordinary in relation to the curfew in Kashmir, it is important to stress that the ordinary in Kashmir is subsumed in violence, making it in itself a different kind of ordinary. As Kelly (2008) has argued in the context of Palestine, there is no reference point for the ordinary; this is also the case in Kashmir, as this is the kind of life people have lived for many decades now, leaving no reference point for the 'ordinary' from their own context. And yet, there is a desire for a different life that would have nothing to do with violence and where the ordinary could really be so. Here, the curfew is very much a part of the ordinary because of its consistently repeating nature. Whenever a curfew is declared, people synchronize their ordinary routines around it – for example, people have already stocked up on food or other essentials such as medicine. This also means that the ordinary entails the normalization of violence (see Kelly 2008). Now, what makes the curfew like that of 2016 in Kashmir extraordinary is its temporality: firstly, owing to its uncertain imposition for nearly two months, with tension around the continuity of ordinary routines; and, secondly, the interlocutors' narration of experiences and practices as being before, during, and after the curfew of 2016, making it quite extraordinary.

It is interesting to address how curfews have different meanings based on the differential lived experiences of inhabitants in the neighbourhood. The media discourses usually conjure up images of empty streets with concertina wires, and stationary or patrolling troops keeping a check on movement, while protestors or stone-throwers try to defy the curfew. However, this section, based on the narratives of the curfew of 2016, explicates the relationship between violence, temporality, and sociality. Through the everyday lived realities, it focuses on the interaction between ordinary and extraordinary experiences via the breakdown and reclamation of routines by the way of strengthening relations of solidarity.

Neighbourhood relations in Nowhatta mean that each household considers some four to five households on each side of their house as their neighbours with whom they maintain constant ties of sharing and trust. Whole families also maintain relations with other families that live further away, if one of their members is befriended by someone in the other household. This constellation can be understood as a great multiplicity of neighbourhoods within the spatially singular Nowhatta.

Neighbourhood ties are mostly maintained by women who meet in the lanes while carrying the trash out of their homes, or accompanying their children to school buses, or going to the market. If not on the main road, one can see women talking in the lanes by their homes. Men usually maintain ties through participation in congregational prayers, where they can catch up with other neighbourhood men.

Some of the neighbours are members of an extended kin group, as well, and the ties in such cases are even stronger. Neighbours are also very important during events like weddings, deaths, illness, injuries, arrests, or other celebrations such as the passing of matriculation (10th class) exams. Dependency on neighbours increases during these times. When physical distance or episodes of violence make it impossible to see extended family and kin group members living further away, neighbours take on a larger role. Neighbouring women fulfil important roles of caring for their neighbours' households and guests at weddings and times of bereavement. These practices must be reciprocated in due time, with the same affection and willingness, while failure to do so leads to what is called *malal* (hurt) in Kashmiri. Neighbourhood relations are also justified on the basis of Islam. People would often talk about the rights of neighbours and their duties towards them, which must be fulfilled to please God.

Nowhatta lacks space and houses are located very close to each other but even so most of the houses maintain a garden and some even have a kitchen garden. Safia cleared her garden to make it into a kitchen garden after the long curfew of 2016. She now grows tomatoes, *saag* (Kashmiri collard greens), green chillies, and maize there. Whenever I went to her home she would send me off with some corn cobs and a bag of tomatoes. This kitchen garden is as a result of the curfew of 2016 when people had difficulty fetching perishable products like fruit, vegetables, mutton, milk, medicines, and petrol. However, Kashmiris traditionally store essential non-perishable foods like rice, lentils, potatoes, and dried vegetables to last them the gruesome winters, when snow blocks the roads making travel, purchases, and exchanges difficult. Kashmiris' winter provisions thus came in handy during the long curfew, helping them to get through the scarcity and hardships that arose.

Safia makes sure that she sends the treats from her new kitchen garden to her close neighbours with whom she shares an amicable bond and who reciprocate with other food items – for example, some sweet dish or other delicacies. 'Now we don't have to worry about *bandhs*, aunty has made all the arrangements for us', says Aafiya, one of my key interlocutors in the field.

During the curfew of 2016, the neighbourhood suffered greatly in terms of not being able to get perishable food items, medicines, and petrol. Aafiya recalls how her father would go to a place near Dal Lake called Jogi Lanker (which is 4–5 kilometres away from Nowhatta) to get vegetables and petrol.

> Every alternate day he would go at 5 o'clock in the morning with uncle Shafi – another elderly neighbour, to get the required stuff. Although they had a tough time, the old men were scared to send youngsters, like uncle Shafi's sons, out so early in the morning, because if they would have gotten caught by the army for no reason, then it would have been a problem.

Here hierarchy works in a different way – the elderly in some cases have a better chance of getting away as compared to youngsters who are more vulnerable to arrests, beatings, and harassment. Although the distinction between private and public is blurred in terms of safety in Kashmir, parents still often restrict young men from going out very early in the morning or during late night hours as they could easily fall prey to arrests or even death at the hands of state forces.

There are many other examples of neighbours helping each other during the curfew of 2016. When schools were shut, some parents worried about their children's studies. Neighbourhood teachers who worked in different local schools came to their aid and voluntarily taught the children during that time. Some of the children later continued taking private lessons from the neighbourhood teachers even after the long curfew was over.

The curfews are often accompanied by an internet ban or e-curfew, so it is now common in Kashmir to hear the question: Has internet service been suspended? (*internet chukh-e band kourmut?*). The dependency on the internet in conflict is huge, with every family having at least one internet connection via which news of the violence and resistance in any part of Kashmir reaches them. The internet thus keeps people well informed on state violence, while also providing a means to reflect and resist through posts, likes, comments, and shares leading to a formation of a virtual collective. The internet is also used for other purposes like shopping, filling in online forms, watching movies, music videos, chatting, learning new recipes and crafts, etc. When the internet is shut down, which entails blocking all the mobile internet services including the most popular connection provided by a private company, Reliance Jio (popular due to its low prices), there is a frenzied atmosphere in the neighbourhood. Recalling the previous year's experiences, as per a young male interlocutor:

> We were handicapped without internet for 6 months. Imagine being day and night glued to it and then finally when it goes off, it is so hard to manage especially when the political situation is so bad and you do not know how many are dead, injured or blinded.

The internet was cut off many times during my fieldwork; in fact, there was no mobile internet upon my arrival in the field. Internet access was available only via the broadband services provided by BSNL (Bharat Sanchar Nigam Limited), which is a state-owned telecommunications company that provides services to most government offices. Over the course of time, I figured out that a few households in the neighbourhood had such a broadband connection too, and to my surprise many people in the neighbourhood knew about these households. It was during the curfew of 2016 that these households had become so famous that their passwords travelled to the surrounding homes via word of mouth, exhausting their data allowance prematurely. One of the interlocutors, a post-graduate student at the University of Kashmir told me:

> Last year many people came to know about me because I had BSNL broad-band. I gave the password to my friends and some neighbours who approached. However, on some days I would see people who live close by but with whom my family does not have close relations wandering around my home with their mobile phones and it was later I realized that they had been using my internet. But what could you say anyway – people were helpless, some had urgency to fill forms etc. so how could you shun them?

However, this year I constantly keep changing the password because now I believe people should look for other alternatives. Last year was unexpected but by now people should be prepared for this. This year I heard a few more households have installed the broadband connection.

During the curfew of 2016, when the mobile internet connection was completely cut off, the neighbourhood helped people to pass the time and keep busy. Some of the men would go to each other's house to play cards, badminton, cricket, etc. These get-togethers for games seldom happen on ordinary days, but this long curfew transformed the neighbourhood by creating these new spaces of sociality. In August 2016, when it was Saika's birthday, she had pestered her parents that she wanted to celebrate it despite the curfew. Knowing how difficult she is, Rehana, her mother, had baked a cake at home and invited a few neighbourhood children for a *kahwa* party (*kahwa* is a type of flavoured Kashmiri tea which is brewed for a long time and has added saffron and almonds, and is usually made on special occasions). According to Rehana, 'kids were accompanied by elders keen on joining the get-together. I had invited just the kids but some of the elders came in as well and the evening passed in chatting and having more and more *kahwa*.'

Normally elders do not attend children's get-togethers. If they did come to drop their children off, they would turn away from the main door even if you insisted that they come inside. However, during the long curfew some of them would come in hoping that no one would mind because of the *bandh*. During ordinary or non-curfew times, people hardly ever spend time with neighbours, their routines restricted by work and family, but this curfew paved the way for new ways of coming together or new socialities as a way of coping and passing time.

Bourdieu has talked about social capital as the aggregate of the actual or potential resources linked to the possession of a durable network of more or less institutionalized relationships of mutual acquaintance and recognition (Bourdieu 1986, 248). This network of relationships is the product of investment strategies, individual or collective, consciously or unconsciously aimed at establishing or reproducing social relationships and transforming contingent relationships into relationships that are at once necessary and elective, implying durable obligations subjectively felt (Bourdieu 1986, 249). In the neighbourhood, social capital seems to work wonders during the days of curfew when neighbours reach out to each other. Such solidarities not only sustain them amidst crisis, but also lead to the generation of other forms of capital like a Wi-Fi password or food from a neighbour's kitchen garden, etc. And, as Bourdieu (1986, 249) argues, 'such exchange transforms the things exchanged into signs of recognition that reproduce the group'.

From the narratives of the inhabitants it is quite evident that, even when the curfew was declared suddenly and people were caught unawares, they looked for ways of coping within the neighbourhood. Over time the practices and patterns of the 'curfewed' life, involving an inward interaction within the neighbourhood resting on solidarities and new socialities, were in some cases routinized for their easy accessibility and to prevent similar discomfort/tensions in case such long curfews were declared again. Therefore, we can see the continuity of the ordinary during the extraordinary, i.e. people trying to maintain their ordinary routines during the extraordinary times of long curfew made possible by the use of social capital. This then led to the continuity of some forms of curfewed-life during non-curfewed days as well, blurring the boundaries between ordinary and extraordinary times. It shows that the violence of the curfew not only causes disruptions and constraints in the everyday, but also leads to possibilities and potentialities for forging solidarities and new socialities.

Neighbourhood – the fractures within

So far the chapter has reflected on how the neighbourhood spatially and materially paves the way for resistance, adding to the archives of public memory and collective history. It further explored how neighbourhood practices and relations in the everyday are accentuated during the violence of the longer curfew. When curfew disrupts the ordinary routine, relations grounded in solidarity both lead to a reclaiming of the ordinary in extraordinary times and, at times, to an adoption of the patterns of the extraordinary in ordinary times.

Further, borrowing from Bourdieu (1986), being a part of the neighbourhood entitles its members to social capital that involves an exchange of different sorts which is subjectively felt as well. However, it would also not be correct to assume that the experiences and practices of the neighbourhood are universal or homogenous, with no ruptures and discordances in terms of suffering, resistance, sociality, or survival. Depending upon the positionality or situatedness of the inhabitants, political violence impacts and shapes people's subjectivities and identities in different ways, in which they are constantly negotiating the boundaries between collective and individual subjectivity vis-à-vis the state of occupation. Maroof, who lives close to the *chowk*, thinks of stone-throwing and the consistent tear gas shelling as hindering his right to a healthy life. The deteriorating health condition of his family members, especially his wife, due to teargas smoke makes him feel helpless and he says:

> I do not pelt stones, those who do run away and do not have to live with this smoke on a daily basis. I do not even go out at that time, and then why do I have to bear the brunt of tear gas smoke? Why is my family sub-jected to this inhuman treatment? Why do we choke and spend money on medicines all year round?

Although Maroof has good ties with the neighbours who live somewhat in the interior of the neighbourhood, who are not affected by teargas smoke as gravely as his family is, social capital also comes with its limitations. No one would take Maroof's family in on an everyday basis whenever teargas shelling starts. 'My own home would now become my grave', Maroof says.

In another case of a middle-class household, not being able to have green tea or not being able to go for a morning walk during curfews leads to what Mumtaz calls unfavourable health. She remembers how she was laughed at when she had asked her husband to arrange for a pack of green tea from somewhere during the curfew of 2016:

> I've been used to having green tea for years now, I go for a walk and then have green tea and it keeps my blood pressure and weight in check; during last year's curfew when my pack of green tea was exhausted, no one would take me seriously when I would ask them to buy some; they would say people do not have food, people are dying and you need green tea, but that was my food and my medicine but they would just taunt me.

The prevailing violence also impinges badly on the daily labourers and vegetable/fruit sellers of the neighbourhood for whom losing a day's earning can be a serious concern. Jalib sells fritters in the neighbourhood and, narrating his experiences during the curfew of 2016, he says: 'it made life too difficult for me'. Jalib walks with a limp and lives with his wife and a child; and with the continuous curfew he sometimes wasn't able to meet his family's basic needs.

> Those 2 months of curfew were the harshest times when we would worry about what to eat for meals and how to arrange for child's milk. A few times during the curfew they provided some 'deals' but what can you buy when you have not been earning anything? Deals/relaxations work for the moneyed but not for the poor like me for whom 'deals' become useless without money! I then had to borrow money, neighbours helped with food sometimes but it was 2 months which is not a joke! Such times are actually a reality check of your helplessness as a poor man! I did not have any savings nor was I a government employee who receives fixed money on a monthly basis despite the *bandhs*.

Here Jalib and others who are not well off, despite the urgency to work to earn their daily bread and to meet their basic needs, are bound to follow the curfew norms and, at other times, the *Hurriyat* calendar. This is a calendar issued by the resistant leadership (*Hurriyat*) aimed at fighting for Kashmir's independence and directing people to different kinds of protests like a shutdown/*hartal* or a march (*chalo*) against state violence.

Many inhabitants recall a 2011 incident in which a local shopkeeper was beaten by protestors in the same neighbourhood for keeping his shop open and defying their dictates to close the shop during stone-throwing. They beat him and he died from his head injuries at a hospital some days later. The locals from the neighbourhood say that those involved in his killing were not from the same neighbourhood, but it made everyone in the neighbourhood very fearful. So, whenever there is a *hartal* and most importantly if stone-throwing is happening, it means one has to conform to the rules as a community and participate – adhering to the collective sentiment and collective resistance to mark your presence as a political community.

To understand the dynamics between collective and individual, solidarities and discords, and continuities and ruptures in the neighbourhood community, I'm borrowing from Arif's (2017) work on the *afterlives of violent events*, where she draws on the work of Esposito to explain the disruptions within community. Esposito argues that individual or personal biographies must be given as a *munus* (i.e. a gift that one gives without an expectation of return) to the collective for the community to exist. This means that private life must be subjugated to public life and the inability to do so due to individual subjective biographies can lead to what is called community disentanglement (Arif 2017, 132). Taking further Arif's argument and her usage of a 'we' versus 'them' and 'we' versus 'I' framework to talk about community and its disruptions (p. 127), I attempt to explain the simultaneous existence of collective and individual subjectivities in a Kashmiri neighbourhood. Here, 'we' is the Kashmiri neighbourhood community with a shared subjectivity and collective identity following on from collective suffering and collective resistance against 'them', which is the state and its appendages like the military, paramilitary or the police, etc. This 'we' versus 'them' might allow us to talk about a homogeneous political subjectivity of the neighbourhood community in particular and of the Kashmiri community in general. However, it is in this 'we' versus 'I' where, when the shared political subjectivity is measured against the individual subjectivity such as that of Maroof, Mumtaz, or Jalib, we notice the fractures within the community, underscoring the presence not only of a collective/coherent shared political subjectivity, but also personal or individual/ incoherent subjectivities.

One's subjectivity is not fashioned in isolation but rather the presence of the 'other' is quintessential to its formation. While the state remains the overarching other for the Kashmiris, within the local community there are also different 'others' based on the different ways state violence impacts the inhabitants, bringing out the tensions between 'we' (symbolizing a collective against 'them') and 'I'. The 'I' here of Maroof, Mumtaz, and Jalib suffer in different ways, and yet their individual subjectivities are subsumed under the collective 'we'. Although they are aware of their singular subjectivity as they narrate their personal experiences of suffering and anguish, they endure it via what Arif (2017) describes as turning reciprocal living on its head, giving away their singular subjectivity to be a part of the collective 'we'.

The discussion on discords and discontinuities also invites attention to the collective political subjectivity vis-à-vis the subjectivity of being human. In the course of my fieldwork, on the eve of *shab-e-qadar*,[2] a local Kashmiri policeman was lynched to death by some men outside the *Jamia Masjid*. According to the locals, the policeman had been involved in some suspicious activities like recording the sermon of the chief preacher, who is also one of the main resistance leaders, and taking photographs of the men attending the sermon. When inquiries were made about his behaviour, he had taken out a gun and shot two men, leading other men to attack him and lynch him to death.

Kashmiris, based on their collective political subjectivity and collective identity shaped by resistance are considered a threat to the state. However, if you are a Kashmiri and working for/with the police or 'security' agencies then you are siding with the state and aiding it in control and oppression, thus becoming a threat to other Kashmiris. As conflict and its strategies work both above and underground – producing a pervasive, generalized fear as to who could be the potential threat – borrowing from Bhan (2013), this ever-present suspicion endangers human security for the sake of national security.

After the lynching event, it became hard for people in the neighbourhood to take sides. One could see the confusion and helplessness on the faces of the people with regards to the incident. Ironically everyone talked about the ethics and values of 'being human', or perhaps the lack of them. Those defending the lynching supported it because the policeman had been a collaborator with the state, 'not human enough' because he worked with/for the oppressors against his fellow Kashmiris. On the other hand, those condemning the lynching also did so on the grounds of 'being human', stating that he was after all a fellow Kashmiri and a pawn in the hands of the state who did not deserve to be attacked and killed in this way, elucidating different and contradictory meanings of ethical human behaviour within the same neighbourhood, breaking the homogeneity and monolithic understanding of the neighbourhood community at yet another level.

Conclusion

The attempts to recover a sense of the everyday are a central part of the process by which people live through and beyond violence (Das 2007). It is not the transcendent that allows people to live amidst devastation, but the banal (Das 2007). This chapter has explored the everyday of a Kashmiri neighbourhood through the persistence of ordinary in the extraordinary times underlining the dynamics of subjectivity, agency, and survival.

The exploration of everyday through the temporalities of the curfew shows how usual curfews are a part of the ordinary, and how during the extraordinary times of the long curfew of 2016 the neighbourhood turned inwards, sustaining itself via the forces of social capital. The neighbourhood both stands cut off from the rest of the city and yet devises ways to sustain and continue with ordinary routines via solidarities and new spaces of sociality. It reclaims the ordinary in extraordinary times – in a way normalizing the extraordinary, while there is also a continuity of the patterns and practices of extraordinary times during ordinary times, thus blurring the lines between 'ordinary' and 'extraordinary'.

The chapter further problematizes the homogeneity of the collective political subjectivity that assumes everyone suffers and survives in the same way. Depending upon their positionalities and the way violence impacts the inhabitants, when individual subjectivity is measured against collective political subjectivity, i.e. when 'I' is measured vis-à-vis 'we', it shows the disruptions and ruptures within the community. And yet, these singular subjectivities are given away, as a gift (Arif 2017), for the sake of the collective, which is unable to adjust to or accommodate these individual or personal subjectivities. The permeable boundaries of the neighbourhood community, which forge a collective by giving space to resistance against any event of violence outside the neighbourhood, become impermeable when 'I' is measured against 'we' and where 'I' must be given up or personal sufferings endured, underscoring the negotiated and shifting boundaries of the neighbourhood community.

An exploration of everyday life and routines in Nowhatta thus offers a different lens to make sense of the violence. Instead of normatively moving from violence to the everyday, the chapter underscores the mundane and banal not only as the sites of violence, but as avenues for negotiation as well. It becomes more about the disruptions and continuities of the everyday mapped through the relations between ordinary and extraordinary times, and the shifting boundaries of collective and personal subjectivity imbuing every day with the power to make and remake itself.

Notes

1 Pseudonyms have been used for interlocutors throughout the chapter to maintain anonymity.
2 Also known as Layat-ul-Qadr or the Night of Power, as it is one of those nights in the last ten days of the month of *Ramadan* (Islamic month when Muslims fast) when the Quran was revealed to the Prophet Muhammad, and Muslims pray and recite the Quran the entire night in the mosques or at home.

References

Anjum, Aaliya and SaibaVarma. 2010. 'Curfewed in Kashmir: Voices from the Valley.' *Economic and Political Weekly* 45 (35): 10–14.

Arif, Yasmeen. 2017. *Life, Emergent: The Social in the Afterlives of Violence*. Hyderabad: Orient Blackswan.

Bhan, Mona. 2013. *Counterinsurgency, Democracy, and the Politics of Identity in India: From Warfare to Welfare?* London: Routledge.

Bose, Sumantra. 2011. *Kashmir: Roots of Conflict, Paths to Peace*. Cambridge: Harvard University Press.

Bourdieu, Pierre. 1986. 'The Forms of Capital.' In *Handbook of Theory and Research for the Sociology of Education*, edited by John. G.Richardson, 241–258. New York: Greenwood.

Brubaker, Rogers. 2002. 'Ethnicity Without Groups.' *European Journal of Sociology/Archives Européennes de Sociologie* 43 (2): 163–189.

Chamlee-Wright, Emily and Virgil Henry Storr. 2011. 'Social Capital as Collective Narratives and Post-Disaster Community Recovery.' *The Sociological Review* 59 (2): 266–282.

Das, Veena. 2007. *Life and Words: Violence and the Descent into the Ordinary*. Berkeley: University of California Press.

Duschinski, Haley. 2009. 'Destiny Effects: Militarization, State Power, and Punitive Containment in Kashmir Valley.' *Anthropological Quarterly* 82 (3): 691–717.

Duschinski, Haley, Mona Bhan, Ather Zia, and Cynthia Mahmood, eds. 2018. *Resisting Occupation in Kashmir*. Philadelphia, PA: University of Pennsylvania Press.

Ibrahim, Farhana. 2018. 'Wedding Videos and the City: Neighbourhood, Affect and Community in the Aftermaths of the Gujarat Earthquake of 2001.' *South Asia: Journal of South Asian Studies* 41 (1): 121–136.

Jasani, Rubina. 2008. 'Violence, Reconstruction and Islamic Reform—Stories from the Muslim "Ghetto".' *Modern Asian Studies* 42 (2–3): 431–456.

Junaid, Mohamad. 2019. 'Counter-Maps of the Ordinary: Occupation, Subjectivity, and Walking Under Curfew in Kashmir.' *Identities*: 1–19.

Kazi, Seema. 2010. *In Kashmir: Gender, Militarization, and the Modern Nation-State*. New York: South End Press.

Kelly, Tobias. 2008. 'The Attractions of Accountancy: Living an Ordinary Life During the Second Palestinian Intifada.' *Ethnography* 9 (3): 351–376.

Lofranco, Zaira. 2017. 'Negotiating "Neighbourliness" in Sarajevo Apartment Blocks.' In *Migrating Borders and Moving Times: Temporality and the Crossing of Borders in Europe*, edited by Donan Hashtings, Madeleine Hurd, and Carolin Leutloff-Grandits, 26–42. Manchester: Manchester University Press.

Peteet, Julie. 1996. 'The Writing on the Walls: The Graffiti of the Intifada.' *Cultural Anthropology* 11 (2): 139–159.

Ring, Laura A. 2006. *Zenana: Everyday Peace in a Karachi Apartment Building*. Bloomington, IN: Indiana University Press.

Robinson, Rowena. 2013. 'Naata, Nyaya: Friendship and/or Justice on the Border. Frontiers of Embedded Muslim Communities in India.' In *Boundaries of Religion: Essays on Christianity, Ethnic Conflict, and Violence*, edited by Rowena Robinson, 242–261. New Delhi: Oxford University Press.

Said, Edward. 1983. 'Opponents, Audiences, Constituencies and Community.' In *The Politics of Interpretations*, edited by William J.T. Mitchell, 7–32. Chicago, IL: University of Chicago Press.

Schofield, Victoria. 2000. *Kashmir in Conflict: India, Pakistan and the Unending War*. London: IB Tauris.

Sorabji, Cornelia. 2008. 'Bosnian Neighbourhoods Revisited: Tolerance, Commitment and Komšiluk in Sarajevo.' In *On the Margins of Religion*, edited by Frances Pine and Joao De Pina-Cabral, 97–112. New York: Berghahn Books.

9 Syrian migration and logics of alterity in an Istanbul neighbourhood

Hilal Alkan

Since 2013, Istanbul has been going through a drastic demographic and material change due to migration. However, this is certainly not a change that is foreign to the city. Migration has always been one of the major forces that has created, shaped, and reshaped Istanbul. This time, the force has been set into motion by the Syrian revolution and the violence that followed. While Turkey opened its borders to Syrians fleeing the war and chose not to employ encampment as the main policy tool, Istanbul became a major hub and settlement destination for Syrians, whose numbers exceed 3.5 million all around the country. Those with residences registered in Istanbul numbered around 550,000 as of June 2019 (Göç İdaresi Başkanlığı 2019), but the number residing in Istanbul is estimated to be much higher. The mass arrival of Syrian migrants and the fast pace of demographic change have posed an 'imminent, embodied and affective challenge' (Gökarıksel and Secor 2018) to residents of Istanbul at varying degrees and intensities.

This chapter develops a spatialised understanding of how this challenge is met, with a focus on the logic of alterity that is at play within daily encounters between Turkish nationals and self-settled city-dwelling Syrian migrants in an Istanbul neighbourhood. The elements under scrutiny here are embodied logics, not always verbally articulated and often missing the coherence of a rational argument. Such embodied logics also constitute a negotiation in and of an urban spatiality: Which bodies can live where and under which conditions? Hence, the chapter illustrates the formation of a neighbourhood by long-term residents and newly arrived Syrians acting out and acting upon differences.

This, however, is not a question of inclusion and exclusion. While some Istanbul neighbourhoods have absorbed more migrants and others fewer, putting the question in a binary formulation leaves out conditional acceptance, shifting positions, and long-term relations. In order to develop a more dialogical and relational—hence inclusive—frame, I resort to Engin Isin's (2002) conceptualisation of citizenship as an ongoing negotiation with alterity and apply it to the smaller scale of a particular lower-class neighbourhood, Kazım Karabekir. Following Isin, in Kazım Karabekir, I explore the encounters and relations between Turkish and Syrian nationals that are shaped and tainted by a spectrum of strategies that build on difference.

According to Engin Isin, in any polity, between and within social groups, 'logics of alterity embody differentiation and distinction, not only as strategies of exclusion' (2002, 25), but also as strategies of solidarity, competition/contestation, and hostility. These strategies are quite fluid and are not exclusive of one another. Depending on the dominant political narratives, power of the social actors, and primacy of alignments, they are employed even by the same people against those who are seen as different. Hence the distinction between them is explanatory, yet not exhaustive; and they are only to be understood in their relationality. Broadly speaking, solidaristic strategies consist of affiliation, identification, care, and alliance, while competitive or agonistic strategies refer to contestation, resistance, and tension. The latter often arise when citizenly privileges or the definition of citizenship (hence belonging and entitlement) are felt to be under threat. Finally, antagonistic or hostile strategies include estrangement, exclusion, expulsion, and oppression (Isin 2002, 32). Their distinctiveness lies in the fact that they announce a refusal to relate and interact, although relationality precedes and conditions them. At the neighbourhood level, these strategies not only have discursive effects, but are also very much materialised in the space and on bodies.

Kazım Karabekir, therefore, is not simply the setting where hostilities or solidarities are enacted. Neighbourhood, as the location of a 'thrown-togetherness' (Massey 2005), is the precondition of these strategies. Physical proximity is, if not the cause, certainly a catalyst of various social and contagious emotions; as well as the strategies to contain or express them. At the neighbourhood level, these strategies are enacted through daily transactions and within mundane encounters. They are often embedded within the texture of ordinary life, but this does not make them less significant, particularly in the lives of newly settled Syrian migrants. Their livelihoods are strongly tied to the propensity of their neighbours to enact certain strategies. The Kazım Karabekir neighbourhood is also being formed and re-formed through these mundane affairs. The outlook of the neighbourhood is changing, public space is being reinvented, neighbourly relations are shifting, and new rifts are coming into being. Therefore, the neighbourhood space is nothing like it was ten years ago. This is a formative moment, laden with anxieties, as well as possibilities.

However, a word of caution is necessary here: This is a rather dynamic field. As the parties involved in the violent conflict in Syria (including Turkey) constantly change positions, form alliances, and make new military and political decisions; as the economic crisis in Turkey deepens; and as the government discourse on Syrians is volatile; predominant and legitimate positions continuously shift. Therefore, the ethnographic examples presented here should be taken as snapshots of ever-moving social actors and ever-changing configurations. Closely dependent on the political emotions in circulation (Ahmed 2004) and the affective atmosphere (Anderson 2009) of the time period, one or another of the three logics predominate. It is also important to recognise that these strategies fall along a spectrum, where they are not neatly separated from each other.

Before moving on with the ethnographic accounts of the strategies that turn anxieties into possibilities and prospects into catastrophes in the hands of hostile, competing, and welcoming neighbours, I will outline the structural factors that delimit the strategies available to all parties involved and introduce the neighbourhood.

Syrian migrants in Turkey

Since the Syrian uprising began in 2011, Turkey has been receiving mass migration from Syria. Turkey's open-door policy in the first years of the conflict and its geographical proximity to heavy-conflict areas made it a natural destination for more than half of the Syrians seeking refuge abroad. Turkey initially responded to the migration wave as a humanitarian crisis that would not last long—a prediction that went well beyond wishful thinking to active military involvement. Within this approach, camps were established and standard relief procedures were put in place. However, encampment was quickly sidelined as the sole response and Syrians were allowed to travel and settle all around the country. Hence, even in their heyday, camps only accommodated 15% of the Syrian migrants. Since the end of 2016, all camps have gradually been closing.

Syrian migration forced the Turkish state apparatus to face many challenges caused by its migration legislation, which was inadequate to provide an immediate response to the needs of fleeing Syrians. Turkey is party to the 1951 United Nations Convention on the Status of Refugees. However, it still maintains a geographical limitation, which in practice results in refugee status not being granted to anyone not coming from Europe. Still, the number of non-Syrian asylum seekers in the country exceeds half a million. These migrants from different parts of the Middle East, Asia, and Africa enter the country in unregulated ways and some apply to the United Nations High Commissioner for Refugees (UNHCR) to be resettled in a third country. While they wait, they are given a temporary status and are assigned to designated towns for residence. Those who do not register with the UNHCR, or those who cannot survive in those towns, go unregistered (Biehl 2015).

Syrian migrants, on the other hand, have received different treatment from the start. The border was kept open for them and their primary address has never been the UNHCR, as very few sought permanent resettlement in a third country. Until 2014, Syrians in Turkey remained in a legal limbo, although there have been many practical arrangements to address their everyday needs, like access to healthcare and education. Turkey's new law on foreigners and international protection came into force in 2014. Syrian migrants were then officially recognised under a legal category created specifically for them: temporary protection. Under this new classification, Syrian refugees were given ID numbers and their social rights were better delineated. Turkey's temporary protection scheme does not define a limited period for the protection to continue, nor does it chart what comes after. It also separates Syrians as a unique category falling outside of constitutional and international protection (Kıvılcım 2016; Baban, Ilcan, and Rygiel 2017). So, although the classification means legal recognition, it also leaves Syrian nationals in Turkey vulnerable to shifts in national and international politics, as shown by the purge that took place in the summer of 2019 that led to the deportation and mass detention of thousands of Syrian nationals (We Want to Live Together Initiative 2019).

However, since 2013 there have also been many regulations aiming to improve migrants' welfare. Registered Syrians have access to healthcare services and public education. Yet, these rights are subject to travel restrictions. If a Syrian migrant under temporary protection leaves the town where she is registered without permission (and because certain cities like Istanbul do not accept new registrations, moving with permission is often not possible), her social rights are cropped. She would also face the risk of deportation, as the events of summer 2019 showed. Legally, Syrians under temporary protection can apply for work permits, although in practice a great portion of Syrian employment is in the informal sector, as the permits are virtually impossible to get. So far only 32,000 permits have been issued, while the number of Syrians in the labour force is estimated to be over one million (Kirişçi and Uysal Kolasın 2019).

In terms of relief, assistance, and advocacy, the efforts have been abundant yet scattered. The Turkish state directs camp-based relief efforts and has selectively co-operated with international non-governmental organisations and local civil society. However, for self-settled Syrians, support coming from central state institutions has been very limited. Civil society organisations and municipalities working in the field have partly compensated for this lack (Mackreath and Sağnıç 2017; Danış and Nazlı 2019). Especially during the early years, many already-existing charitable organisations (religiously motivated or not) also assumed responsibility and shifted their operations, or expanded them to cover Syrians. Moreover, informal aid initiatives flourished everywhere around the country with or without prospects of formalisation. As the years passed and Syrians' needs changed, some of these informal or ad-hoc initiatives dissolved, while others turned into registered non-governmental organisations (NGOs). Their material capacities and command over resources have drastically changed with access to international funding. In so far as they secured funds and became local partners, they professionalised (Mackreath and Sağnıç 2017).

There has also been a lot of effort at the neighbourhood level, all around the country, initiated by people who wanted to reach out to their new Syrian neighbours. They are the most miniscule of the networks out there (compared to giant international organisations or national NGOs) and the least formalised or institutionalised. But they are also the ones that have produced the most fertile circumstances for the potential creation of personal connections between newer and older residents. Because, unlike the case of many other civil society organisations, those who devote their time and resources to providing aid and services in these neighbourhood networks reside in the same place as those who are the object of their compassion, pity, or fraternal solidarity. This cohabitation of space allows for daily and habitual encounters to take place and long-term relationships to flourish. What colours these relationships don and which modalities they assume are discussed below.

Syrian migrants and the changing Istanbul neighbourhoods

Syrian migrants are dispersed across Istanbul, but are more concentrated in some districts. These are often the districts where the lower-middle classes and the urban poor reside (Erdoğan 2017). Some have larger Kurdish populations with their own histories of forced migration; the Kurds of Syria settle in those areas more easily (Kılıçaslan 2016; Kaya 2016). Others are known for their Islamist politics; families of combatants feel safer there. Some neighbourhoods have textile workshops in every second building; Syrian migrants settle nearby to find low-paying jobs in Turkey's informal economy (Erdoğan 2017). Some migrants follow their previously settled kin when choosing between Istanbul neighbourhoods. Around the city, many neighbourhoods are being demolished and rebuilt (via hotly contested urban transformation laws), so the evacuated buildings provide shelter to those who are expelled elsewhere: the Yazidis and the Roma (Dom) of Syria and Iraq (Cox 2016; Kalkınma Atölyesi 2016). In wealthier middle-class districts, affluent Syrians, who came right after the conflict began, live in their own flats (Kaya 2016).

The main pattern, however, is that Syrian migrants have settled in districts that are defined by higher poverty levels, religious-conservative attitudes, and more reliance on solidarity in the social environment (Erdoğan 2017, 31). Most of these preferred districts are located on the European side of Istanbul, where small- and medium-scale industry is also concentrated. In some neighbourhoods of these districts (like Zeytinburnu), the Syrian presence is readily and immediately felt, as the official population ratio has reached over 9%. Yet, in the middle-class residential district of Ataşehir on the Asian side, a completely new development of high-rises, Syrians amount to less than 0.03% of the residents (Erdoğan 2017). My research site is in close proximity to Ataşehir, yet representing and embodying a whole different universe.

Kazım Karabekir is a neighbourhood of Ümraniye, a large district on the Asian side of Istanbul. Like the overall district of Ümraniye itself (Erder 1996), Kazım Karabekir is a neighbourhood that came into existence through internal migration to Istanbul in the 1970s and 1980s. The building stock mostly consists of apartment blocks that gradually replaced the earlier single- or double-storey illegal shanty houses, while the owners slowly moved up the social ladder from poor immigrants to lower-middle-class landlords (Işık and Pınarcıklıoğlu 2001). Still, there are plenty of these earlier single-storey, detached houses with sub-standard infrastructure on the outskirts of the neighbourhood where it has chipped away woods.

In Ümraniye, Syrian nationals constitute approximately 2.5% of the population (Erdoğan 2017). In Kazım Karabekir the ratio is slightly higher, given the fact that it is relatively cheap compared to some other, more central neighbourhoods in the district. Syrians are dispersed throughout Kazım Karabekir, though most live in the gloomiest and most unkempt of rental apartments, neighbouring lower-class Turks and Kurds, almost all of whom have their own histories of migration. These similarities in background and in current living conditions, however, do not guarantee sympathy. On the contrary, they sometimes fuel competition and animosity. Throughout the rest of the chapter, I will first illustrate solidaristic and welcoming strategies of some residents of the neighbourhood in response to the arrival of Syrian migrants. Then, I will move on to discuss agonistic/competitive and antagonistic/hostile responses that have been rising since.

Aiding neighbours

Between late 2015 and 2017, I conducted ethnographic research in Kazım Karabekir on neighbourhood initiatives aiding Syrian migrants. The network I worked with was organised around an elderly couple, Aliye and Ismail, who have been living in the neighbourhood since its earliest days as an informal settlement on the urban periphery. Their first encounter with Syrian migrants is indicative of 'thrown-togetherness' (Massey 2005) as a key characteristic of urban contexts. Aliye is in her sixties and lives in a four-storey apartment building that her family had built in place of their single-storey shanty house 20 years ago. The building accommodates two of her sons and their families, and one other rented flat owned by Aliye. Shops and storage rooms occupy the ground floor. One day in 2013, Aliye noticed new tenants moving into a similar storage facility in the apartment block across the road. She tried to figure out who these people were and eventually gathered up a few neighbours and her courage to knock on their door.

With the help of her rudimentary Arabic and investigative eye, Aliye figured out that the family was from Aleppo, consisting of a mother, her five daughters, two sons, and a granddaughter. They did not have any furniture, any appliances, nor even a stove to heat up the cold, damp basement flat. In the following days Aliye, her husband Ismail, and the neighbours gathered the necessary supplies and brought them to the flat. Then, quickly, they were introduced to other Syrian migrants in the neighbourhood, who were either relatives or simply acquaintances of the first family. The network grew larger every day. Needs were immediate: coal stoves to survive the winter, duvets, carpets, cooking stoves, and fridges. So Aliye and Ismail reached out to their neighbours, friends, and relatives. Ismail contacted local business owners asking them to donate items or to sell them at considerable discount. They gathered new and second-hand items to furnish the flats and arranged for their delivery.

Aliye and Ismail's network consisted of more than 300 Syrian households and a lesser number of Turkish nationals. Through this network, they drew resources from companies, NGOs, municipalities, and religious groups, but also considerable contributions from unaffiliated individuals. Their network, like many other ad-hoc, informal neighbourhood-based initiatives, mediated flows of cash, food, clothing, coal, furniture, household items, jobs, bill payments, medicines, and information on bureaucracy, healthcare, schooling, transport, and accommodation.

During the first few months they had to try hard to reach potential donors. However, as the plight of Syrians received more and more media coverage and sympathy, those who wanted to help approached them without solicitation. It was a period in which 'hospitality' and 'religious brotherhood' were the key terms of the governmental and public discourse (Carpi and Pınar Şenoğuz 2019; Alkan, forthcoming). Through word of mouth, awareness of their efforts grew and they started to attract larger resources as well as more demands upon these resources. At the time I came to know them, their contacts included people from top business circles, but most were still their neighbours and acquaintances. All the Syrian migrants in their network resided in and around the neighbourhood. A few of them became more central as gatekeepers who aided newcomers in many ways, and their recommendation carried more weight in the eyes of Aliye, Ismail, and others involved.

The importance of this network is twofold: First, it provides vital services and meets the immediate needs of newcomer Syrian migrants. Second, and more importantly for this chapter, these encounters often spark long-term relationships between Syrian and Turkish nationals. Aliye, Ismail, and other actively involved residents of Kazım Karabekir not only mediate aid distribution, but also have neighbourly relations with some of the Syrian migrants they have met through this network. They now exchange visits with each other, take part in celebrations and in grief, and care about each other's well-being. They have become ordinary neighbours, with neighbourly relations.

Here some explanation is needed to clarify what neighbourliness often means in Turkey. Neighbourliness is *gendered care labour*. It is labour in the sense that it requires deliberate effort to sustain, and it requires physical and emotional work. It carries many elements of care work, although it is much more overtly reciprocal than most dependency work. Still, it loosely falls into the category of what Eva Feder Kittay (1999) calls 'love's labour'. And as such, it is strongly gendered. It creates different obligations and entitlements for men and women, and is often sustained by the daily interactions of women (Özbay 2014). For men, the location of neighbourliness is often public spaces, while for women neighbourliness opens the insides of homes to each other. It creates intimacy. In a 2006 nationwide survey, 85% of the participants reported that they have frequent contact with their neighbours, a number much higher than contact with immediate family members (Özbay 2014). For the women of Kazım Karabekir, too, neighbourly relations entail daily (and often intimate) communication, visiting each other without invitation, and coming together on special occasions like prayers after funerals, religious festivals, and engagement ceremonies, among many others. They also occasionally involve looking after each other's children and helping with seasonal house and kitchen work (Ayata and Güneş-Ayata 1996). In that sense, what I mean by neighbourliness in this chapter goes far beyond civil urban behaviour—i.e. sharing a building and a greeting in the morning. Most of my Syrian interlocutors have a few such *neighbourly* Turkish neighbours, although not enough for their taste.

Neighbourly relations, as mutual relations by definition, do not entail a one-way flow of goods and care. They are not one-off aid relations. They stretch over time and require reciprocity. Syrian migrants, as neighbours, actively give and contribute to the relationship that started with the assistance they received from the semi-organised efforts of the neighbourhood network. My research shows that giving to neighbours has a catalysing effect in creating long-term intimate relations between people who share an urban locality (Alkan, forthcoming). In order to understand this power of giving in creating cohesion, it is useful to look into the basic principles that characterise most gift relationships. What follows is a brief summary informed by Marcel Mauss (1990), Pierre Bourdieu (1997), Jonathan Parry (1986), Maurice Godelier (1999), and Annette Weiner (1992).

Every gift begets a return. It does not mean that the person who gives wants or expects a return. Rather, if the giver expected a return when giving, it would not count as a genuine gift, but a calculated transaction. What makes a thing or a gesture a gift is the notion that it is given without an expectation of a return. However, gifts still oblige a return and leave a burden on the shoulders of the receiver. Reciprocity releases the burden, although it does not entail exact equivalence and it does not clear the balance. Instead of providing a closure, it elicits more gifts because the counter gift is a gift itself, begetting a return: this is the gift cycle as Marcel Mauss (1990) formulated it. With every new gift the cycle resists completion; it rather goes up as a spiral in time.

This is the dynamic that runs through the neighbourhood network in Kazım Karabekir, sometimes to the pain of its participants. Unlike municipalities or large NGOs entering a neighbourhood and distributing food packages or coal every once in a while, the neighbourhood networks that were formed around the Syrian migrants created longer-term relationships. In these networks, there are intimate relations to be formed through the circulation of gifts, whether material or immaterial, such as compassion, love, and the very physical activities of providing care. In the case of close neighbourly relations, circulation of gifts does not have the power to equalise positions; rather, because the positions of givers and receivers are interchangeable, they offer escapes from the tight grip of superiority and inferiority. Exceeding street-level politeness, these relations involve normalisation—normal here signifying ordinary neighbourly care and intimacy, expected between people residing in proximity to one another within an urban context.

However, intimate care is not necessarily a bed of roses. It entails expectations, responsibilities, and enactment of power. Care can be a burden, as much as a proliferating gift (de La Bellacasa 2017). It is immanently and immediately linked with discipline, and this aspect of care is prominent in the lives of the Syrian women I was acquainted with in Kazım Karabekir. Hasna, for example, told me at great length how annoyed she was about the comments of her neighbours on her standard of cleaning. They, she said, would not find her home clean enough whatever she did. When she told me that, my non-vocalised hasty reaction was, 'that is none of their business'. However, Hasna's reaction was something else. She was not particularly disturbed by the fact that her neighbours saw it their right to comment on the cleanliness of her flat. Her protest was against their ignorance about different ways of cleaning—the habits she developed while cleaning her single-storey, earth-brick village home near Aleppo. For her, sweeping the house would not even count as cleaning; she was used to washing the floors. Yet, she did not have this chance in Istanbul. So she had to settle for a practice she despised and was not quite proficient at. Slowly, she said, she learnt to do it as her neighbours suggested, although quite half-heartedly.

If cleaning is one area of discipline neighbours enact on each other, another is the education of children. The Turkish state responded to the educational needs of Syrian children in two ways. At first, it encouraged the establishment of temporary education centres where Syrian refugee children would be educated by Syrian teachers using a Syrian curriculum (Aras and Yasun 2016). The Syrian community, international NGOs, and various non-governmental actors from Gulf countries funded these schools. The Turkish Ministry of Education often provided infrastructure, such as buildings (Aras and Yasun 2016). In these schools, the language of education was Arabic, although students took a few hours of Turkish language classes every week. In 2018, there were 229,000 children enrolled in these facilities (MEB 2018). In line with the growing awareness that most of the Syrian migrants would stay in Turkey, at the end of 2017 state policy shifted towards public schooling. However, at the time of my research, there were two of these centres, popularly known as Syrian schools, in the vicinity of Kazım Karabekir. The issue was a source of constant concern among the Turkish nationals and a matter of intervention in their relations with their Syrian neighbours. Parents who sent their children to Syrian schools were kindly questioned or warned about the children's lagging Turkish skills.[1] Parents of publicly educated Syrian children, on the other hand, were worried about the fact that their children had already lost their fluency in Arabic or Kurdish. Yet, the neighbours dutifully and almost militantly argued for education in Turkish, even at the price of losing native languages, and celebrated these children's quickness in learning Turkish.

Neighbourly interventions are not limited to children's education or cleanliness. They extend to a variety of areas such as gender relations, piety, consumption, and dress codes. Yet they all flow within the context of neighbourly care and concern. Such concern for the well-being of their Syrian neighbours—well-being assessed on a normative scale—constitutes a significant part of the 'grammar of good intentions' (Ryan 2003) that is spoken in Kazım Karabekir. Coming from caring neighbours, whose good intentions are presumed, the migrants often tolerate such disciplinary attempts. They are seen as part of the care package that constitutes mutuality between older and newer residents and performs vital functions in their tenuous existence as war refugees. And, perhaps more importantly, they may tolerate this because Syrian migrants regularly and increasingly come across much harsher strategies, which enact a logic of alterity, than these solidaristic ones.

Competing neighbours

In the summer of 2016, I travelled around Kazım Karabekir for a full day with Aliye and two other women in a car driven by Ismail. One of the women was Serap, whose family owned a leading discount supermarket chain that has branches in every middle- and lower-income neighbourhood around the country. The other was Meliha, Serap's assistant in her philanthropic activities. In Meliha's bag, the supermarket vouchers waited to be distributed to Syrian migrants in the neighbourhood. Aliye and Ismail had worked out the list of households to be visited the day before. Meliha asked them to decide who they should be given to, and Serap would decide how many. So, Aliye and Ismail considered those with immediate needs, the people they knew better, and whom they felt obliged to give to due to their personal closeness. They weighed the options, decided who could wait one more week, and put those onto a second list. Instead of full names and addresses, both of the lists consisted of illegible jots to remind them of the people. The rest was left to Aliye's extraordinary memory and Ismail's impressive navigation skills.

That day we visited 15 households, each one of them known to Ismail and Aliye through previous contact. Aliye had closer relations with some. She told their stories to Meliha and Serap, kissed their kids, exchanged gossip with women in Turkish and Arabic, and gave tips about this and that. Others were barely known to her. On these occasions, Aliye and Ismail sometimes became a little disoriented while finding the exact flat. While waiting for them, we would stand on the pavement, looking for a clue. Yet, every time we stopped the car and looked around to locate the flat, we would also be looked at by the neighbours. Some would only stare inquisitively, while others would openly ask who we were looking for. The next question would then be: 'Are you distributing something? You always and only come for the Syrians, don't you?'

Most would stop there, with this expression of resentment. They would then shrug and show us the flat: 'some Syrians live here, but I don't know the name'. Nevertheless, on one occasion, a woman who was simply passing by and figured what we were up to, did not stop at a sullen expression of resentment. Instead, she held us captive on the side of the road and told her own story of migration, domestic violence, and extreme poverty. Ismail tried to stop her, either because he did not want Meliha and Serap to get annoyed and leave; or he did not want to stay on the street any longer, with the fear that many others might come, hearing the way she pled. Maybe both. The woman, however, would not accept interference. Aliye looked embarrassed and tried to explain how some residents of the neighbourhood were particularly cheeky. The woman responded to that with fury: 'Aren't we poor enough for you? What do

I have that they did not have?' Meliha soothed her eventually and she left with a voucher in her hand, promising that she would not tell anyone.

The sense of being unjustly left out of the compassion directed to Syrian migrants is not unique to this one woman. A similar feeling of resentment is often to be found at the neighbourhood level in Kazım Karabekir, in line with nationwide sentiment. Turkish nationals often complain that the Syrian migrants are treated differentially, while their needs that arise from being in similarly poor circumstances go unaddressed. The complaints often very quickly turn into racist and xenophobic banter, as exemplified by the social media campaigns that take place with increasing frequency (see, for example, #suriyelileriistemiyoruz on Twitter). In general, they are expressed along a spectrum, ranging from mild envy to full-blown misplaced fury, and always point to competition over resources and social opportunities. While in some areas of life this resentment has a more material basis, in most cases it is built upon hearsay and imagination.

A striking example of how far imaginary competition can go comes from a nationwide survey conducted by Murat Erdogan and his team. In 2017, they asked 2,089 Turkish citizens a range of questions aiming to understand their attitudes towards Syrian migrants. When asked about the Syrians' sources of their livelihoods in Turkey, 86% of respondents said that Syrians' basic source of income was the monetary support they received from the Turkish state (2018, 65). Yet, only 22% of the 886 self-settled Syrian nationals who took part in the research said that they had received support (in cash or in kind; from individuals, NGOs, or any other agent) in the previous year; and less than half of this 22% received support from the Turkish Red Cross (the only category that is directly related to the state) (2018, 120). Given that Red Cross financial support is completely funded by the EU (Reliefweb 2019), it is possible to say that not a single self-settled Syrian declared having received financial support from the Turkish state in 2016.

This mind-blowing mismatch between the reality of Syrian migrants and the imagination of Turkish nationals is symptomatic of a phenomenon that is not restricted to Turkey. In an age of rising right-wing populisms, such phantasmagorical misperceptions are the bread and butter of xenophobia (see, for example, Yılmaz 2012). Wendy Brown diagnoses in this a displacement. Borrowing the concept of *ressentiment* from Friedrich Nietzsche and building on Freud's work on narcissism, she approaches resentment as a displacement of one's own suffering on an object (Brown 1993, 2017). Whether or not that object has anything to do with the suffering itself—i.e. the question of causality—is deferred. What matters are the wound and the easiness of the target at a given point in time and space. Hegemonic political discourse, as well as impunity for the crimes committed against those who are targeted, produce the necessary (un)truth effect. Efforts to re-place responsibility upon the actual actors—onto landlords and property firms for tremendously increasing their rents in order to benefit from the vulnerability of migrants; on business owners, for exploiting migrant labour; on the government, for entrenching the conflict in Syria … etc.—do not even make a dent on this conviction. Turkish nationals position themselves in competition with Syrian migrants and ask for a restoration of their privileges.

At the neighbourhood level in Kazım Karabekir, complaints that refer to competition cluster around rents and health services. As both Syrian migrants and Turkish nationals have drastically experienced, rents have at least doubled over the last six years in Istanbul. The increase is often attributed to Syrian migration and migrants report that they are harassed on the street with a personalised version of this accusation (see also DW 2019). A much bigger increase, however, has taken place in property sales prices. Financial analysts point to explosive market behaviour similar to that of the housing bubble that led to the stock market crash in the USA in 2007–2009 (Cagli 2019). Yet such analysis does not change the fact that rent increases have far exceeded increases in income. Syrians are targeted more easily than the construction firms and hedge-funders, as they are more proximate, more visible, and certainly much more powerless.

Being in the vicinity of a major public hospital, Kazım Karabekir residents have long been privileged in their transport-free access to healthcare. 'After the Syrians came, the hospital became too crowded', they now complain. However, their greatest concern is not the overcrowding of the hospital, which has always been the case (as they also admit after a bit of scrutiny), but the fact that Syrian migrants were receiving free (yet limited) healthcare. Syrians under temporary protection are also exempt from the contribution payments Turkish nationals have to make to receive certain services or medication. In the eyes of Kazım Karabekir residents, this signifies turning citizenship on its head. Losing their taken-for-granted privileges feels like an affront to their citizenly identity.

While the competition is often on the terms of social citizenship, there are times when debates about the nature and inclusiveness of political citizenship also flare. In July 2016, President Erdoğan announced gradual naturalisation for highly qualified Syrian migrants.[2] The announcement immediately caused heated debate, particularly around the fact that naturalisation would give new citizens the right to vote. An MP from the main opposition party issued a parliamentary question asking whether Syrians were being given citizenship to increase votes for the governing party in the upcoming elections. In street interviews, Turkish nationals expressed feeling hurt by the possibility of sharing their 'most important citizenly right' with the migrants. In the majority view, Syrian migrants were not qualified for citizenship for multiple reasons, some more frivolous than others, ranging from a lack of character to being noisy (BBC 2016). Yet, public outrage was visible and well-reasoned: regardless of who the Syrian migrants would vote for, voting was not something the Turkish nationals would accept sharing, as it was considered the main signifier of citizenship.

Going back to Engin Isin's (2002) conception of citizenship as a dialogical construction, what we notice here is a dispute over the perceived privileges that are expected to come with formal citizenship status. Among the people of Kazım Karabekir, what citizenship should mean is discussed in relation to the

Syrian migrants who, by falling outside of formal citizenship but still being part of the polity, can be positioned as *the other* to be (sometimes violently) competed with. Competition over public resources, entitlements, rights, and statuses mark Turkish citizens' views of Syrian migrants, despite the fact that the migrants have very little power and leverage in this competition. Whether it is the meagre services provided by volunteers or access to state-provided services like healthcare and education, what Syrian migrants are entitled to is scrutinised and quickly loses its connection to reality. In this fantasy realm, Syrian youth enter universities without examinations, every Syrian household receives monthly cash benefits from the state, and they will soon be given public housing. Such narratives, Sara Ahmed says, 'generate a subject that is endangered by imagined others whose proximity threatens not only to take something away from the subject (jobs, security, wealth), but to take the place of the subject' (2004, 43). This subject then moves as far away from his new neighbours as physical limitations permit—agony turns to antagonism.

Hostile neighbours

It should have been apparent by now that Kazım Karabekir is not solely marked by convivial neighbourly relations and inclusive strategies. Turkish nationals who see a threat in the arrival of Syrians often express their frustration in agonistic terms, as seen above. However, there is also a very strong (and increasingly strengthening) vein of reaction, which covers a whole range of hostile, antagonistic strategies. They vary from daily insults on the streets to driving Syrians out by means of constant harassment.

All my Syrian interlocutors had a lot to say about the hostility they had to endure in the neighbourhood. They told me about hostile neighbours who complain about their children playing in front of the buildings using racist terminology; xenophobic comments at the weekly market; reproachful stares on the minibuses; and all sorts of open insults. A public park is particularly identified with such hostility. Almost every Syrian woman I met in the neighbourhood had a story to tell about the tiny neighbourhood park, which is the only recreational space they have access to. The park is a favourite location for women to take their children to the playground and relax under the trees. It is also loved as a picnic location at the weekends, despite its small size. However, it is also the place where Turkish youth regularly harass Syrians. Some women even witnessed a beating, which then went unreported.

Hostility is not a uniform behaviour. It is expressed in a variety of ways, some subtle, some overt. It is also not uniformly distributed. Syrian migrants become its targets differentially, depending on their age, gender, religious observance, and ethnicity; again, in a dialogical relation with the agents of antagonistic strategies. While hate is often expressed in gendered and sexualised ways (Gökarıksel and Secor 2018, 8), ethnicity is also pertinent to its flow and direction, sometimes shifting the positions in the established 'organisation of hate' (Ahmed 2004).

Dalia has lived in Kazım Karabekir with her husband and four children (two biological and two adopted) for the past 18 months. She is an Arab from Hasakah (Haseke/Hesiçe), the region where most of the Syrian Kurds of the neighbourhood came from. Before coming to Istanbul in 2018, she lived in Urfa, a Turkish city on the Syrian border, and gave birth to both of her sons there. They now live in a single-storey shanty house in the heart of the neighbourhood. The house has a large garden with plenty of fruit trees. If the weather is good, we always sit on Dalia's dark red sofa under the walnut tree. If the weather does not permit, however, we are stuck in one of the two cold and damp rooms. None of us like that. Dalia complains; I nod understandingly with regret about my helplessness to change the situation; the children make their siblings pay for their boredom.

Dalia is bored too. She dreads the winter, but summer is not particularly bright either. Alongside all her other troubles, she suffers greatly from loneliness. In a city of 16 million, including half a million Syrian migrants, she feels completely isolated. Her violent husband leaves home every morning to collect and sell recyclables. Two of the children go to school, and Dalia is left all alone with two little boys. Every time I visit her, she tells me that I am the only one who knocks on her door. I ask her about the neighbours. Her face drops. 'No neighbours' she says. When I keep asking, she shrugs: 'All neighbours are Kurds'. Given the context, for her, this is self-explanatory.

Since August 2016, the Turkish Armed Forces have been actively increasing their territorial control over Northern Syria/Western Kurdistan (Rojava) in alliance with the Free Syrian Army and various armed factions. In the summer of 2019, when I last visited Dalia and asked her about her neighbours, negotiations with the United States about the eastward expansion of Turkish military control towards Kobane and Hasakah were ongoing. A crackdown on the Kurds, living on the northern side of the border, in Turkey, was already under way. Later that month, the Turkish Ministry of Interior once again removed the elected mayors of three Kurdish majority towns from office and started another purge of the Kurdish political movement in Turkey. The ministry also began encouraging Syrian Arabs to voluntarily return to Syria to be resettled in the areas controlled by the Turkish Armed Forces and their allies in the Kurdish territories. Deportations and forced 'voluntary' returns were also documented. Within the entangled politics of Turkey and Syria, the Kurds of Kazım Karabekir, who migrated from the southeast of the country, neighbouring Syria, had many reasons to see Arabs coming from Rojava (including Kobane, Hasakah, and Afrin) as allies of the evil. What Dalia sees as self-explanatory in her Kurdish neighbours' deliberate distance is indeed a function of history and politics.

Derya Özkaya's chapter in this volume (Chapter 10) illustrates how the Syrian war is reterritorialised in an Istanbul neighbourhood, as residents of the neighbourhood literally became parties in the conflict. In Kazım Karabekir, the war is brought home, not by Turkish nationals joining military forces in Syria, but through the embodied presence of Syrian migrants. And yet, that war was never far away. Even before active military occupation, Turkey was intimately involved with the uprising and conflict in Syria (Phillips 2016) and this involvement was always closely linked to Turkey's own problems with Kurds (Taştekin 2017).

If Dalia were a Kurd from Hasakah, how would living in an Istanbul neighbourhood be for her? My other interlocutors answer this hypothetical question. Anti-Kurdish sentiment constitutes a major element of the affective nationalist repertoire in Turkey (see Zeydanlıoğlu 2008; Saraçoğlu 2009). Syrian Kurds, however, are doubly stigmatised in most parts of the country. They carry the burden of being 'the Syrians', as well as being the 'wrong' Syrians. The hostility they receive is exacerbated by their categorical inclusion in the 'wrong' side of the conflict. My Kurdish research participants report being repeatedly rejected by landlords, being threatened on the street, and insulted for being Kurds. However, they also say that volunteers from the neighbourhood network—Ismail and Aliye in particular—would not discriminate according to ethnicity or religion. Yet, this does not mean that they felt safe from the start. When I asked about the demographics of the Syrian migrants in Kazım Karabekir, a woman from the neighbourhood network said that she did not know any Kurdish Syrians but proffered that, if she were to come across some, they would likely not disclose their ethnicity because they would be afraid of being considered 'terrorists'.

This fear, sympathetically diagnosed by a volunteer who is not a politically engaged solidarity activist, but a pious housewife, is even more prevalent in the lives of migrants with fewer allies in Turkey's rifted political landscape. For similar reasons, Turkey has not been a favoured destination for Syria's Christians (Arsu 2016; Kreidie 2017). For the Yezidis of Syria, who had to take refuge in Turkey after the genocidal attacks of ISIS in 2014, passing as Muslims is a vital strategy to counter the hostility they would otherwise face in Turkish cities. Already being recognised as 'out of place' (Douglas 1970), they are not only considered strangers but also as dangerously dirty. When I asked how they approach the Yazidis, Zeynep, the co-ordinator of a local foundation's relief operations, was perplexed. She couldn't think of any Yazidis, neither in the neighbourhood nor in Umraniye. However, she later recalled an uncanny encounter in another district, during which she had felt threatened and unsafe due to the behaviour of the men surrounding her. She left that place hastily, clinging to her bag, and after that encounter she made only home visits accompanied by a male colleague. 'They must have been Yazidis', she concluded.

'They', from what Zeynep described and to my knowledge, could not have been Yazidis. Yet being seen as the strangest of the strangers to have arrived with the Syrian migration, for her, Yazidi was the most suitable category to put these 'uncanny' men in. For her, Yazidis represent the most alien, the furthest away in social distance, and their proximity creates the greatest anxiety. Their presence feels like a threat to her feeling at home while carefully and successfully navigating alterity in her encounters with Syrian migrants. This is the territory she feels too unsafe to navigate.

Conclusion

Throughout this chapter, I have illustrated different navigational strategies that the inhabitants of Kazım Karabekir employ in their interactions with their new Syrian neighbours. I approached these strategies as solidaristic, agonistic, and antagonistic, feeding into and taking root in established societal tensions in Turkey. With no claim to exclusiveness, I laid out the spectrum on which these strategies are situated, starting with neighbourliness (and its flipside) and ending with hostility, all involving recognising Syrian migrants as strangers. Zeynep's encounter with the assumed Yazidi group marks the limit of recognition here, and thus a limit to the willingness to engage.

Through these engagements and disengagements, the neighbourhood itself is being formed. Its identity is changing as conviviality is debated, not always deliberately and in verbal articulation, but through the mundane affairs of daily life in the neighbourhood. Proximity creates tensions, but it also creates intimacy. Some residents welcome this change, even if hesitantly, while others loathe it. In any case, the logics of alterity that play out change what Kazım Karabekir is. The notion of 'the neighbourhood' is too slippery at the moment; it is 'an unpredictable amalgam of the familiar and the unknown' (Herzfeld 1991, 91). In the end, it is a place that citizens and strangers share without really knowing how to share it. And sharing means both cohabiting and dividing.

I finish this chapter with a brief note on strangers. Strangers, in Sara Ahmed's (2000) words, are not those who do not fit onto the map, into the order. Instead, what makes them strangers—Syrians or refugees—is the fact that they fit very well into mappings of alterity. The map is alive, like those interactive maps in which parts of the globe swell or shrink depending on the variable you choose to see. Syrians may overpopulate the corner of human misery in one click; while in another they are the desperate yet sly competitors in a tight job market. However, there is nothing new about these categories, variables, criteria, and adjectives used to recognise and describe Syrian migrants. They have all existed before, used for others, kept well in stock. They presume recognition at first sight, because Syrian migrants are already recognised as *the Syrians*. Agonistic and antagonistic strategies make robust use of this established repertoire. Solidaristic strategies, on the other hand, may (and only may) have an element of surprise, learning, and innovation—i.e. changing the map. Paternalism, pity, multicultural welcoming, and humanitarian motivations all have their own pre-markings, but if they open the way for long-term relations, they carry this potential. Neighbours have many colours. Neighbourliness, however, has only shades.

Notes

1 It should be noted that the anxiety around these schools also involved the religious and political upbringing of the children, as many Turkish nationals were afraid that the schools were the craddle of Salafi militants. This concern was shared by Ismail, who was a self-declared Islamist and a pious man himself.

196 *Hilal Alkan*

2 So far the numbers have been miniscule in comparison to the total number of Syrian migrants. As of August 2019, 92,000 Syrians have acquired Turkish citizenship, half of which are children (Euronews 2019).

References

Ahmed, Sara. 2000. *Strange encounters: Embodied others in post-coloniality*. New York and London: Routledge.
Ahmed, Sara. 2004. *The cultural politics of emotion*. New York and London: Routledge.
Alkan, Hilal. (forthcoming). 'Gift of hospitality and (un)welcoming Syrian migrants in Turkey'. *American Ethnologist*.
Anderson, Ben. 2009. 'Affective atmospheres'. *Emotion, Space and Society* 2 (2): 77–81.
Aras, Bülent Aras, and Salih Yasun. 2016. *The educational opportunities and challenges of Syrian refugee students in Turkey: Temporary education centers and beyond*. Istanbul: IPC-Mercator Policy Brief.
Arsu, Şebnem. 2016. 'No Christmas for Syrian Christians in Turkey'. *Politico*, 24 December. Accessed September 9, 2019. www.politico.eu/article/no-christmas-for-syrian-christians-in-turkey/.
Ayata, Sencer, and Güneş-Ayata Ayşe. 1996. *Konut, komşuluk ve kent kültürü*. Ankara: TC Başbakanlık Toplu Konut İdaresi Başbakanlığı.
Baban, Feyzi, Suzan Ilcan, and Kim Rygiel. 2017. 'Syrian refugees in Turkey: Pathways to precarity, differential inclusion, and negotiated citizenship rights'. *Journal of Ethnic and Migration Studies* 43 (1): 41–57.
Başkanlığı, Göç İdaresi. 2019. Accessed September 11, 2019. www.goc.gov.tr/gecici-koruma5638.
BBC. 2016. 'Suriyelilere vatandaşlık: AKP seçmeni ne diyor?' 13 July. Accessed September 12, 2019. www.bbc.com/turkce/haberler-turkiye-36780544.
Biehl, Kristen Sarah. 2015. 'Governing through uncertainty: Experiences of being a refugee in Turkey as a country for temporary asylum'. *Social Analysis* 59 (1): 57–75.
Bourdieu, Pierre. 1997. 'Marginalia-some additional notes on the gift'. In *The logic of the gift: Toward an ethic of generosity*, edited by A.D. Schrift, 231–243. London and New York: Routledge.
Brown, Wendy. 1993. 'Wounded attachments'. *Political Theory* 21 (3): 390–410.
Brown, Wendy. 2017. 'Apocalyptic populism'. Published in German translation. *Blaetter für Deutsche und Internationale Politik* (8): 46–60, in English. Accessed August 26, 2019. www.eurozine.com/apocalyptic-populism/.
Cagli, Efe Caglar. 2019. 'Explosive behavior in the real estate market of Turkey'. *Borsa Istanbul Review* 19 (3): 258–263.
Carpi, Estella, and H. Pınar Şenoğuz. 2019. 'Refugee hospitality in Lebanon and Turkey. On making "The Other"'. *International Migration* 57 (2): 126–142.
Cox, David. 2016. 'Syria's gypsy refugees find sanctuary in an Istanbul ghetto'. *The Guardian*, June 2, 2016, sec. cities. Accessed September 11, 2019. www.theguardian.com/cities/2016/jun/02/syrias-gypsy-refugees-sanctuary-istanbul-turkey-ghetto-how-long-will-last.
de La Bellacasa, Maria Puig. 2017. *Matters of care: Speculative ethics in more than human worlds*. Minneapolis: University of Minnesota Press.
Douglas, Mary. 1970. *Purity and danger: An analysis of concepts of pollution and taboo*. Middlesex: Pelican Books.

DW. 2019. 'Türkiye'de yaşayan Suriyelilerin ayrımcılıkla mücadelesi'. 20 June. Accessed September 10, 2019. www.dw.com/tr/t%C3%BCrkiyede-ya%C5% 9Fayan-suriyelilerin-ayr%C4%B1mc%C4%B1l%C4%B1kla-m%C3%BCcadelesi/ a-49265942.

Erder, Sema. 1996. *İstanbul'a bir kent kondu: Ümraniye*. Istanbul: İletişim Yayınları.

Erdoğan, M. Murat. 2017. *Urban refugees from "Detachment" to "Harmonization": Syrian refugees and process management of municipalities: The case of Istanbul*. Istanbul: Marmara Belediyeler Birliği.

Erdoğan, M. Murat. 2018. *Suriyeliler Barometresi: Suriyelilerle Uyum İçinde Yaşamın Çerçevesi*. Istanbul: İstanbul Bilgi Üniversitesi Yayınları.

Euronews. 2019. 'Bakan Soylu: Çocuklar dahil 92 bin Suriyeliye vatandaşlık verildi'. Accessed September 9, 2019. https://tr.euronews.com/2019/08/02/bakan-soylu-92-bin-suriyeliye-vatandaslik-verildi-suleyman-soylu.

Godelier, Maurice. 1999. *The enigma of the gift*. Chicago: University of Chicago Press.

Gökarıksel, Banu, and Anna J. Secor 2018. 'Affective geopolitics: Anxiety, pain, and ethics in the encounter with Syrian refugees in Turkey'. *Environment and Planning C: Politics and Space*.

Herzfeld, Michael. 1991. *A place in history: Social and monumental time in a Cretan town*. Princeton, NJ: Princeton University Press.

Işık, Oğuz, and Melih Pınarcıklıoğlu. 2001. *Nöbetleşe Yoksulluk: Sultanbeyli Örneği*. Istanbul: İletişim Yayınları.

Isin, Engin Fahri. 2002. *Being political: Genealogies of citizenship*. Minneapolis: University of Minnesota Press.

Kalkınma Atölyesi. 2016. *En Alttakiler: Suriyeli Dom Göçmenler Yoksulluk ve Ayrımcılık Arasında Göç Yollarında*. Ankara: Kalkınma Atölyesi: Bilim, Kültür, Eğitim, Araştırma, Uygulama, Üretim ve İşletme Kooperatifi.

Kaya, Ayhan. 2016. *Syrian refugees and cultural intimacy in Istanbul: "I feel safe here"*. RSCAS 2016/59. EUI Working Papers. Florence: European University Institute Robert Schuman Centre for Advanced Studies Global Governance Programme.

Kılıçaslan, Gülay. 2016. 'Forced migration, citizenship, and space: The case of Syrian Kurdish refugees in İstanbul'. *New Perspectives on Turkey* 54: 77–95.

Kirişçi, Kemal, and Gökçe Uysal Kolasın. 2019. 'Syrian refugees in Turkey need better access to formal jobs'. *Brooking Institute Blog* (blog), July 18. Accessed September 12, 2019. www.brookings.edu/blog/order-from-chaos/2019/07/18/syrian-refugees-in-turkey-need-better-access-to-formal-jobs/.

Kittay, Eva Feder. 1999. *Love's labor: Essays on women, equality and dependency*. New York: Routledge.

Kıvılcım, Zeynep. 2016. 'Legal violence against Syrian female refugees in Turkey'. *Feminist Legal Studies* 24 (2): 193–214.

Kreidie, Marwan. 2017. 'Why do so few Syrian Christian refugees register with the United Nations high commissary for Refugees?' *Rozenberg Quarterly*. Accessed September 9, 2019. http://rozenbergquarterly.com/why-do-so-few-christian-syrian-refugees-register-with-the-united-nations-high-commissioner-for-refugees/.

Mackreath, Helen, and Şevin Gülfer Sağnıç. 2017. *Civil society and Syrian refugees in Turkey*. Istanbul: Helsinki Yurttaslar Derneği.

Massey, Doreen. 2005. *For space*. London: Sage.

Mauss, Marcel. 1990. *The gift: The form and reason for exchange in archaic societies with a foreword by Mary Douglas*. London: Routledge Classics.

MEB. 2018. 'Bakan Yılmaz: 608 bin Suriyeli çocuğun eğitime erişimi sağlanmıştır'. Accessed August 27, 2019. www.meb.gov.tr/bakan-yilmaz-608-bin-suriyeli-cocugun-egi time-erisimi-saglanmistir/haber/15549/tr.

Danış, Didem, and Dilara Nazlı. 2019. 'A faithful alliance between the civil society and the state: Actors and mechanisms of accommodating Syrian refugees in Istanbul'. *International Migration* 57 (2): 143–157.

Özbay, Ferhunde. 2014. *Akrabalık ve Komşuluk İlişkileri: Türkiye Aile Yapısı Araştırması, Tespitler, Öneriler*. Ankara: Türkiye Aile ve Sosyal Politikalar Bakanlığı Yayınları.

Parry, Jonathan. 1986. 'The gift, the Indian gift and the "Indian Gift"'. *Man* 21 (3): 453–473.

Phillips, Christopher. 2016. *The battle for Syria: International rivalry in the new Middle East*. New Haven and London: Yale University Press.

Reliefweb. 2019. Accessed September 12, 2019. https://reliefweb.int/report/turkey/emer gency-social-safety-net-essn-helping-refugees-turkey-march-2019.

Ryan, Susan M. 2003. *The grammar of good intentions: Race and the Antebellum Culture of benevolence*. Ithaca: Cornell University Press.

Saraçoğlu, Cenk. 2009. '"Exclusive recognition": The new dimensions of the question of ethnicity and nationalism in Turkey'. *Ethnic and Racial Studies* 32 (4): 640–658.

Taştekin, Fehim. 2017. *Rojava: Kürtlerin Zamanı*. İstanbul: İletişim Yayınları.

We Want to Live Together Initiative. 2019. *Two weeks of deportations*. Press release. August 8, 2019. Istanbul.

Weiner, Annette B. 1992. *Inalienable possessions: The paradox of keeping-while giving*. Berkeley: University of California Press.

Yılmaz, Ferruh. 2012. 'Right-wing hegemony and immigration: How the populist far-right achieved hegemony through the immigration debate in Europe'. *Current sociology* 60 (3): 368–381.

Zeydanlıoğlu, Welat. 2008. 'The white Turkish man's burden: Orientalism, Kemalism and the Kurds in Turkey'. *Neo-Colonial Mentalities in Contemporary Europe* 4 (2): 155–174.

10 Negotiating solidarity and conflict in and beyond the neighbourhood

The case of Gülsuyu-Gülensu, Istanbul[1]

Derya Özkaya[2]

Introduction

'There is no politics [*siyaset*] in Gülsuyu anymore!' said Dursun with a sense of despair as we were waiting for the participants of the meeting. He was not alone; many of my informants repeated the same narrative in our conversations about the current political situation in Gülsuyu-Gülensu, Maltepe's sub-neighbourhood on the Anatolian side of Istanbul. Gülsuyu-Gülensu, as some other working-class neighbourhoods in Istanbul, has always been very quick to react to nationwide political developments, organise political events, and/or take to the streets. As Kemal, another interlocutor in my field research, said, 'Any kind of socio-political development in the country has reverberated here as well.' However, currently, 'Gülsuyu-Gülensu is silent', grumbled Kemal, unhappy about not being able to mobilise more people for the demonstrations.

In February 2017, soon after I started my fieldwork in Istanbul, I started to pay visits to the Gülsuyu-Gülensu neighbourhood as part of my research. While dissent was becoming visible in public spaces again, and the silence had been broken by political campaigns after a long period of retreat from the streets due to the state of emergency imposed in July 2016, there was a dispirited atmosphere in the neighbourhood. Kemal, like the other interlocutors I talked to, was frustrated about the political apathy in the neighbourhood. The local political actors' lack of enthusiasm for organised politics did not match some of the residents' expectations of the political potential of their neighbours. Thus, they repeatedly shared nostalgic and melancholic narratives on the good old days of the neighbourhood during my fieldwork. What they longed for was their 'organised [*örgütlü*] neighbourhood' with intimate neighbourly relations, various political sensibilities, and strong solidarity networks creating a sense of unity. While being an organised neighbourhood was a source of dignity for some of the residents, the price that had to be paid in return created fear and unrest among others. They had witnessed different types of contestations – religious, ethnic, political – since the establishment of the neighbourhood; however, the political sphere of the neighbourhood had become much more complicated due to various developments in recent years.

Neighbourhood as both the physical place of neighbours [*mahalle*] and a set of relations between neighbours [*komşuluk*] works as a scale at which the heterogeneity of its residents coexists with its organised solidarity. Focusing on Gülsuyu-Gülensu as a case study, in this chapter, I will argue that taking neighbourhood as a scale of analysis helps us understand how heterogeneity and homogeneity function together in making the neighbourhood a space for making politics on multiple levels. Benefiting from the scholarship on the social production of space (Smith 1993; Swyngedouw 1997; Massey 1999; Brenner 2004) and the discussions of the local and the global as scales of analysis (Swyngedouw 1998; Peck 2002), I will show how conflicts might create new solidarity networks or dissolve existing ones in different contexts. In the first part of the chapter, I will focus on solidarity organisations and collective political actions against urban transformation projects in Gülsuyu-Gülensu. Within these organising efforts, neighbourhood appears not only as a place of politics; it also enables contentious politics and claim-making (Tilly and Tarrow 2007). In the second part, I turn to what kinds of political contestations are taking place in Gülsuyu-Gülensu, ranging from 'drug wars' to the Syrian war, which indicates not only the political constitution of the neighbourhood, but also how neighbourhood functions as a place for making politics through 'the networks of relations at every scale from local to global' (Massey 1994, 265). Lastly, I will discuss how neighbourhood works as a scale of analysis, a space for claim-making, and a place for making politics.

Elaborating on the complexity of various actors and dynamics in (un)doing politics in the neighbourhood, this chapter asks how the political is spatially (re) produced in the context of local and trans-local conflicts and what happens to solidarity networks when local contradictions coincide with (inter)national conflicts. Sharing a preoccupation with the socially constructed nature of the place (Massey 1994) and focusing on a specific instance of place-making (Lefebvre 1991), this chapter explores the ways in which neighbourly relations are spatially, historically, and politically reproduced through ongoing struggles. Following Massey's call 'to conceptualize space as constructed out of interrelations, as the simultaneous coexistence of social interrelations and interactions at all spatial scales' (1994, 264), it demonstrates how the everyday life of the neighbourhood is increasingly interconnected with socio-political transformations at both national and international levels. Thus, it addresses socio-political transformations through the scale of an urban neighbourhood alongside international developments. Focusing on different practices of claim-making and documenting the everyday violence in the making of a neighbourhood, this chapter reveals the complexities of the making, contesting, and reconstruction of neighbourhoods under processes of socio-economic transformation.

Gülsuyu-Gülensu as a site of coexistence

Gülsuyu-Gülensu neighbourhood (shown in Figure 10.1) on the northern side of Maltepe district in the Asian part of Istanbul was formed as one of the typical squatter [*gecekondu*][3] neighbourhoods as a result of migration from Anatolian

Figure 10.1 Gülsuyu-Gülensu.
Source: Photo by Derya Özkaya.

cities and towns to Istanbul during the 1950s. The immigrants from Erzincan, Sivas, and Dersim (Tunceli) provinces were the first residents of the settlement, with a large majority of Alevis and Kurds. It is characterised by its political tendency toward the left as Alevi, Kurdish, and working-class populations constitute the main body of left-wing politics in Turkey. With the massive political mobilisation and struggles for political and economic rights organised by leftists, socialists, and revolutionaries in the 1970s, Gülsuyu-Gülensu also became one of the *gecekondu* neighbourhoods known by socialist organisations. The inhabitants of the neighbourhood engaged in a land occupation in 1978 that was continued by several migrants in the following years. However, the hegemonic power of the socialists in the neighbourhood was interrupted by the 1980 military coup. The socialists who took part in the construction of the *gecekondu*s and their supporters were imprisoned, and the political activities in the neighbourhood were disrupted until 1987. However, the historically leftist heritage of the neighbourhood has continued to define its political reflex in times of political conflict shaking up society, or during state or police interventions targeting either political actors or ordinary residents trying to protect their houses against demolition attempts (Aslan 2010; Yıldız 2013).

From 1987 onwards, the socialists started to establish local associations with the residents that led to new political mobilisations in the neighbourhood. Both the political organisations and activities organised by the socialists and local political organisations like hometown associations, mukhtar's offices, religious communities, and issue-based local initiatives capable of organising solidarity events created a lively political environment. Due to different waves of migration from the 1990s, Gülsuyu-Gülensu received immigrants from all parts of the country and started to house different cultures, creating a heterogeneous social and political structure in the neighbourhood. While the neighbourhood turned into a multi-ethnic and religious place, its political character was also diversified, with smaller communities mostly based on the inhabitants' 'home-regions' [*hemşehrilik*]. As Kemal explains:

> Although the ethnic and religious conflicts in the neighbourhood [*mahalle*] are minor, regional differences are not really insignificant. I mean, for instance, people from Giresun all live on one side – think of it like a ghetto. People from Samsun are together in one part, people from Bay-burt in the other. All of them are grouped together. So there is not really contact between, say, Kurds [and Turks], or Alevis and Sunni Muslims, as we understand contact and dialogue in that warm and genuine sense, except in some particular spaces. Let's say we're neighbours, and we do what this relationship requires but there's not much of a relationship between this side of the neighbourhood and that side.

The increased number of inhabitants led to the expansion of the neighbourhood and its eventual division, albeit only bureaucratic and administrative. It was divided into two as the Gülsuyu neighbourhood and the Gülensu neighbourhood, with two different local mukhtars. Despite the official regulations, both the inhabitants and the local officers still accept them as one neighbourhood and speak of the Gülsuyu-Gülensu neighbourhood. This culture of living together also enables the establishment of solidarity networks between the two neighbourhoods, especially in the struggle for housing against the urban transformation project. Although the historically leftist heritage of the neighbourhood and the hegemony of the organised leftist struggles are currently spatially limited to the Gülensu neighbourhood, Gülsuyu-Gülensu is accepted as one spatial identity with its politically organised character. This has been symbolised by the neighbourhood's resistance against the urban transformation project since 2004 (Aslan 2010). While the continuous threat and the ongoing struggle against the urban transformation project have been the main political agenda for almost 15 years, the political sphere of the neighbourhood has become much more complicated with the increasing number of political actors, practices, and agendas in recent years.

Since the early 2000s, Istanbul has turned into a huge construction site for capitalist accumulation, neoliberal production, and consumption of space as one of the concrete examples of David Harvey's (2003) conceptualisation of 'accumulation by dispossession'. This period was a turning point for urban renewal unprecedented in speed and scope (Kuyucu and Ünsal 2010; Yalçıntan and Çavuşoğlu 2013; Bartu Candan and Özbay 2014; Çavuşoğlu and Strutz 2014; Kolluoğlu 2014). Triggered by the spatial politics of the Justice and Development Party (AKP)'s neoliberal project, Turkey has been undergoing a spatial redesign through the (re)construction of urban public spaces in the last two decades. The continuous transformation of the city has been fulfilled by an alliance of state, capital, and local governments via displacement and forced evictions, creating urban conflict and resistance at the same time. Some of the *gecekondu* neighbourhoods have transformed silently or with more muted reactions due to conflicts among residents or the AKP's consent-producing policies (Çavuşoğlu and Strutz 2014), and the residents have had to leave their homes. Others, on the other hand, have witnessed collective resistance against the destruction attempts and violent police attacks that have contributed to the development of urban movements in Turkey.

Gülsuyu-Gülensu has also been one of the favoured locations for urban transformation projects since the Istanbul Metropolitan Municipality (IMM) declared an Urban Renewal Plan in September 2004. Although the transformation has not yet begun, it has created unrest among the inhabitants, which has led to a local urban movement. Following the IMM's declaration, the local neighbourhood associations and the two mukhtars took the initiative and organised public meetings, bringing together the inhabitants, local associations, political organisations, professional chambers, local administrations, and researchers. They organised several activities including petitions, collecting letters of objection, and appealing to the court for a stay of execution. This local resistance was supported by large numbers of residents who took part in the local committees, organised and attended public meetings, circulated petitions, and closely followed every detail of the project. Despite their different ethnic, religious, or political backgrounds, the residents of both the Gülsuyu and Gülensu neighbourhoods were able to come together to act. This collective resistance was replicated in different contexts and led to the creation of new solidarity networks such as in the fights against drug gangs and increasing violence in the neighbourhood.

As a result of this collective struggle, the urban transformation project was cancelled in 2012. Following this process, the residents of the neighbourhood started to work on an alternative plan in cooperation with the Chamber of Architects and urban planners, academics, and activists. At the end of this struggle and negotiations with both local and national administrative institutions, the IMM issued its latest development plan on 14 April 2017. With this plan, residents without papers had the opportunity to receive title deeds to the land/ houses where they had been living for decades. This was the starting point of a new phase of the neighbourhood's struggle that aims to ensure that all rights-holders unconditionally benefit from case-law transfers related to suspended plans by preparing a legal infrastructure for citizens able to document that they lived in the area before 2000. The collective production of this alternative plan and its vision to renew the neighbourhood in its own place with its inhabitants make this local struggle unique among urban movements in Turkey.

The unique alternative resistance strategy of the Gülsuyu-Gülensu neighbourhood attracted great attention from many urban activists and researchers, who created an extensive body of literature on the historical and political background of the neighbourhood focusing on local political actors (Özdemir 2013; Yıldız 2013), the impact of the dispossession and displacement policy of the state–

capital alliance (Lovering and Türkmen 2011; Şen 2013), and alternative resistance strategies against the urban transformation project (Şen 2010; Ergin and Tılıç 2013; Özdemir and Eraydın 2017). However, most of the existing studies on resistance in the neighbourhood have a restricted understanding of politics limited only to the struggle against the urban transformation project, while the political sphere of the neighbourhood is much more complex. The Gülsuyu-Gülensu neighbourhood incorporates the multiple roles, actors, relationships, and political subjectivities that have always involved contestation and/or conflicting encounters besides solidarity and togetherness in everyday politics. Thus, there is still a need to unpack the complicated socio-political environment of the neighbourhood, which I will do in the following part by providing a general picture of these conflicting political powers in Gülsuyu-Gülensu.

Gülsuyu-Gülensu as a site of discrepancies

As I was conducting my fieldwork, the first public meeting in response to the new IMM plan took place at a cafe in Gülsuyu with more than 50 men and only 4 women residents. The two mukhtars, representatives of two local associations, a lawyer, and a civil engineer were ready to inform the participants of the details of the plan. Accompanied by non-stop tea service, the organisers of the meeting gave speeches one by one. Nazan, a resident of Gülensu and a lawyer volunteering in the planning process as a member of the municipal council, articulately informed us with a remarkably good command of the subject. After giving detailed information about the planning process and the residents' legal rights, she warned us, emphasising 'we continuously highlight this point: The neighbourhood should not fall victim to politics [*siyaset*]. No matter what, we are neighbours. And we have to execute the neighbourhood conduct above all else,' and concluded her impressive speech by saying that 'our most important weapon is our organisation [*örgütlülük*], our biggest strength is unity [*birlik*]. If we continue to be united, we will not be destroyed by any power [*irade*]. Be sure of this.'

I was impressed by her elocution and I supposed her emphasis on the need for unity and neighbourliness would motivate the neighbours to get involved in the formation of neighbourhood committees; that is, until the first question came from one of the passionate residents of the neighbourhood: Irfan, a worker in his early 30s, was angry at Nazan not because of her speech but because of her political affiliation. Nazan is a member of the Republican People's Party (CHP) and was then working for the municipal council. Irfan, who supports the Justice and Development Party (AKP), implied this opposition, labelling her an 'outsider' despite Nazan living in the same neighbourhood. Suddenly everybody started to grumble, all the voices were intertwined. Ali, the moderator and one of the representatives of the neighbourhood committee, intervened. He tried to calm down the heated atmosphere by saying 'A wolf awaits on every corner when our unity is broken. Colloquially speaking, we would all be the bait', and then added, 'no matter to which political party they belong, those who break us are my enemy'.

After several calls for unity, the residents raised their questions and shared their concerns. Almost all the questions were about their personal and/or familial needs or priorities, such as the current market value of their land, the land register, the number of apartments they will get after the transformation of the neighbourhood, etc. They had similar concerns and/or demands despite the differences in their ideological, religious, and ethnic backgrounds, which make them strategically temporary allies. Although they might resent each other for other reasons, they can come, stay, and act together for the sake of their existence in the neighbourhood. In that sense, neighbourhood appears as a place in which proximity ensures a similar relation to the state-capital power that enables resistance and claims-making in the name of protecting the neighbourhood. What Nazan mentions in her speech, that the neighbourhood must be held higher than any political interest, is an example of such a conception of neighbourhood as a homogeneous place, despite the need for this articulation indicating the complexity of such an assertion. In the end, Nazan's call was not a description of what a neighbourhood is – that is held above politics in itself – but a prescriptive argument for solidarity against the urban renewal project. This is obvious from Irfan's response to Nazan's political affiliation. However, that the neighbourhood is above politics – a statement that seemingly undermines the politics behind the statement itself – shows precisely how a notion of the heterogeneity and homogeneity of the neighbourhood function together in the everyday politics of claim-making.

This meeting motivated me to uncover the political sphere of the neighbourhood, which not only generates the sentiments of unity, solidarity, compassion, etc., but also reproduces discrepancies or inner tensions. Thus, the political in the neighbourhood emerges through, rather than in spite of, these conflictual situations among various political actors in the neighbourhood. While the inner social, cultural, and political dynamics of the neighbourhood itself were already complicated, belonging to the neighbourhood in light of intensified societal and political polarisation has become more difficult. Kadriye, a middle-aged, Alevi woman who has been working in the informal sector for more than 20 years, and one of the active members of the *Cemevi*,[4] elaborated on the issue as such:

> Gülsuyu is a trademark, which represents good and beautiful to us. However, for the middle or upper class, Gülsuyu is a criminal centre, where gangs are created and the radical leftists collide. The system is trying to criminalise neighbourhoods like ours, as if something bad happens here every single day. On the contrary, both the police and the district governorships' sources show that our neighbourhoods are places where the crime rates are the lowest. Because these places are relatively more responsive where residents can react to the problems. However, other districts down the hill are completely insecure.

It is notable how Kadriye feels the need to make a comparison with another neighbourhood – 'districts down the hill' – in order to show how her neighbourhood has remained relatively safe and secure. However, the complicated part of this story is that, in reality, there has been an issue of insecurity in Gülsuyu-Gülensu in recent years, as one of the old residents of the neighbourhood whispered to me, 'People think that Gülsuyu is an organised [*örgütlü*] neighbourhood but it is not any longer. Now Gülsuyu is a neighbourhood where the gangs have swarmed.' However, the roots of such insecurity should be thought to be in line with the intricate relationships between the state–capital alliance and the Turkish state's promotion of violence through oppression of dissent or empowerment of drug gangs in poor urban neighbourhoods. The stigmatisation and criminalisation of organised struggles and political actors in the name of terror in these neighbourhoods has long been one of the main strategies of the Turkish state (Aslan 2004; Cörüt and Evren 2007; Yonucu 2018). Similarly, the increased visibility of various gangs and everyday violence despite immense police control has been given as one of the reasons for the necessity of urban transformation projects since the early 2000s.

Studies on the criminalisation of the urban poor and the sources of urban violence (Wacquant 1999; Comaroff and Comaroff 2016) demonstrate how poverty and persistent social and spatial exclusion generate systematic or organised violence through extra-legal sources of income and power. Although the existing scholarship elaborates on the close ties between non-state armed actors, such as drug gangs or land mafia, and state actors, such as the police and the military (Glebbeek and Koonings 2015; Koonings and Kruijt 2015), there is still a need for further research on how state-sponsored violence in the local setting and the mobility of different violent actors are instrumentalised by the state to promulgate its ideology. Criminalisation of the urban poor and increasing everyday violence, which have cultivated narratives of insecurity and fear in these locations, has always collaborated with de-politicisation of the urban poor by targeting politically organised actors and their supporters (Parenti 2000; Karandinos et al. 2014). At times of political, economic, and social violence, local grass roots movements, solidarity networks, and neighbourhood committees play very important roles in resisting the containment of violence and insecurity through the struggle for collective goods and needs. As most of my interlocutors mentioned, the revolutionary and/or socialist struggles in the neighbourhood that provided safety, security, and solidarity networks have been disrupted by increased police violence in the neighbourhood. The 'dawn raids' as a special form of police operations, with thousands of heavily armed police officers from the Turkish Special Forces, riot police and anti-terrorism units, military tanks and helicopters play an important role in the weakening of the opposition. These police operations and the resistance against them are used to legitimise the need to 'solve the *gecekondu* problem'. While the streets were freed of revolutionaries, the visibility of drug gangs began to increase on the same streets.

Gülsuyu-Gülensu as an arena for violent confrontations

There have always been conflictual encounters between the revolutionary socialists of the neighbourhood and the police in Gülsuyu-Gülensu since the establishment of the neighbourhood. However, the everyday violence was intensified by the violent attacks of drug-trafficking gang members in the neighbourhood during the summer of 2013. Gang members racketeered the artisans, threatened the minibus drivers with the seizure of their vehicles, and beat or shot those who tried to resist. Moreover, many socialists were stabbed by gang members and received death threats. On 6 August, nine members of the Socialist Party of the Oppressed (*Ezilenlerin Sosyalist Partisi*, ESP) who were all Gülsuyu residents were shot in their legs in three attacks. All these attacks occurred under the immense surveillance of both the undercover police and the surveillance cameras at nearly every street corner, but the police never caught the attackers, who were known to everyone in the neighbourhood. This strengthened the suspicion that the drug gangs were being protected and supported by the police.

While it is the young gang members who tend to bear arms in this power struggle, some of the young socialists also advocated armed resistance with the idea of implementing, securing, and dispensing justice in the neighbourhood themselves. Although the militarisation of the neighbourhood youth created a fearful atmosphere among the inhabitants, some of them felt safe being protected by their 'own' people in the 'absence' of state security forces. While the inhabitants were organising marches against the drug gangs, armed socialists were protecting the public against the gang's attacks. This was the starting point of a new phase in the history of the neighbourhood that is referred to as a 'concept change' by Mahir, who has been living in Gülensu since the early 1970s:

> Back then, leftists used to have influence on [drug sellers] since they were not organised. They had organisations but not so powerful. There was not a control mechanism and when organised powers of the left interfered, the problems were solved, but they kept selling. The case is not over because this is a problem created by the system. But we also should have been organised. The more organised the people were, the less power [the drug gangs] had. This was the situation in the past. We were many and we were active, thus they were working underground and doing simpler things. However, after a certain stage of Gezi [the popular uprisings of 2013 in Turkey], they changed the concept and they started to be more organised … They know this job. This is also written in the newspapers. This concept started here and spread through other *gecekondu* settlements.

During a large demonstration on 9 September 2013, the gangs showed up and opened fire on the crowd, wounding several residents and killing Hasan Ferit Gedik. Gedik was living in another working-class neighbourhood, Küçük Armutlu, and he was one of the organisers and active members of neighbourhood assemblies fighting the urban transformation projects in Istanbul. Gedik's political identity and the four bullets in his head left no room for doubt that he was consciously targeted by the gang members. The murder brought the neighbourhood into focus for the rest of Istanbul and the rest of the country.

Following Gedik's funeral,[5] the police launched dawn raids against both the attendees of the funeral and the drug gangs. More than 40 socialists and 22 people assumed to be responsible for numerous crimes, including the murder of Gedik, were taken into custody.

While Gedik's murder created a moment of interruption in the everyday politics of the neighbourhood, it also revealed the fragmentation of political community under the complexity of interwoven actors at multiple levels, and their overlapping political agendas. The following news and developments around the murder displayed the intricate relationships of various political actors at every scale from local to (inter)national and showed that the fear in the neighbourhood was not ungrounded. It was beyond an armed conflict between the marginalised and impoverished young residents fighting for sovereignty in the neighbourhood. During the hearings concerning Gedik's murder, more details about the political identities of the gang members became publicly known. Several news reports showing the gang members' close connections to nationwide mafia leaders and the ultra-nationalist political organisations appeared in newspapers. Besides the journalistic pieces uncovering the connections between the drug gangs and the mafia, the gang members themselves shared their photos and videos on their social media accounts.[6] A quick survey of the internet shows that almost all the gang members who were involved in or supported the violent attacks in the Gülsuyu-Gülensu neighbourhood are also supporters or members of far right, nationalist political parties or associations.

For most of the residents of the neighbourhood, the rationale behind the gang violence was clear: increasing the speculative value of the neighbourhood's land via the displacement of the current residents. Sabri Şakar, the former mukhtar of Gülensu, summarised this connection with these words:

> The price of our land has gone up due to construction work in the sur-rounding areas. Like most of our residents, I, too, believe that the seekers of speculative land value are behind the gangs. We do not want our kids to become drug addicts. And they want to push us out.[7]

Similarly, as Ali Şengül, the president of the Gülsuyu-Gülensu Beautification Association (*Gülsuyu-Gülensu Güzelleştirme Derneği*), expressed during our interview:

> [a]ll the systems in the world do the same: criminalizing and dehumanizing the people and creating a new humankind that is not able to think or ques-tion. Drug trafficking is a serious problem in the neighbourhoods. The police do not sell it directly but they turn a blind eye to the sellers although there are many surveillance cameras here. A simple example: when a citizen makes an illegal demonstration, the police can identify the pro-tester from the shoes he wears but a drug dealer sells drugs under the sur-veillance cameras and nobody opens an investigation about him or does

anything. It shows that the police and the state promote all this. Because the more people become addicted to drugs or alcohol or the more people engage with prostitution or else, it is better for the state. The reasons are never questioned.

Increasing the land value of the neighbourhood through criminalisation and security discourses, and promoting drug trafficking to make the urban poor addicted and incapable of questioning the realities of their lives seem irrelevant. Yet, they have been the government's main strategies of social and political control in poor urban areas. While the rise of everyday violence through the increased visibility of drug gangs cultivates the discourses on the need for securitisation and police patrolling, the urge to 'cleanse' these 'dangerous places' paves the way for the reconstruction of these areas. The association of the urban poor with crime through the reproduction of a fear and security discourse provides legitimate grounds for the implementation of urban transformation projects as the basic forms of gentrification and neoliberal urban policies. Violence – either promoted by the state and state-sponsored actors, or by anti-state forces – often has broader implications for the perceptions of insecurity that pave the way for the reconstruction of neighbourhoods and cities. As Stephen Graham (2011) claims, increased urban violence has led to the diffusion of militarised debates about security, which in turn has brought about increased surveillance and securitisation of urban landscapes with new methods. And all this becomes easier by depoliticising the poor via destruction of solidarity networks and collective resistance through imprisonment of local activists on the basis of their alleged connections to the terror organisations. Thus, the struggle against the criminalisation of the urban poor and the fight against the drug gangs cannot be dissociated from the resistance to the urban transformation project in Gülsuyu-Gülensu. This intertwinement indicates an instance in which the heterogeneity of the neighbourhood, represented by conflicts between drug gangs and the socialist organisations, is produced by a state strategy in order to pave the way for the gentrification of Gülsuyu-Gülensu, against which various political actors organise solidarities. In this case, this dialectical relationship between conflict and solidarity is the result of, and a response to, the long-standing neoliberal urban planning policies that have been executed on a national scale. Yet there are international conflicts that find their expression at a neighbourhood scale in Gülsuyu-Gülensu as well. The Syrian war is one of them.

From 'drug wars' to the Syrian war

Since the beginning of the Syrian war, Turkey has become the largest refugee-hosting country, with a growing population of Syrian refugees who are mostly confined to the margins of society and the poor districts or *gecekondu* settlements where the rent is cheaper. On this point, Gülsuyu-Gülensu differs from other poor urban areas in Istanbul with the insignificant numbers of Syrian refugees living in the neighbourhood.[8] Although the visibility of Syrian refugees is very limited in the neighbourhood, a closer look at the developments there in recent years provides important insights into the direct impacts of the Syrian war.

Following the beginning of the war, poor urban neighbourhoods in Turkey sent many of their residents, mostly youth, to fight for or against the Islamic State (ISIS) in Syria. Not only the fighters, but also their supporters live in these neighbourhoods, and the tension between different parties changes the internal dynamics of the neighbourhoods. However, it is not easy to conduct systematic research on this difficult subject since it can result in targeting or being targeted by the state or state-sponsored gangs that overtly or covertly support the war. Besides, there is no record of the numbers or profiles of the fighters who go to Syria to fight with or against ISIS. Usually, young people disappear without informing anybody and only if they die and their dead bodies are found or recognised can they become publicly known. Other than that, it is something discussed in inner circles but cannot be common knowledge, as I experienced during my fieldwork. However, it is still possible to follow the traces of this 'flow' of fighters at different levels.

Following the Kurdistan Workers' Party (PKK)'s mobilisation to the Syrian border outside Turkey in 2015, the involvement of PKK-affiliated groups in defending Kurds fighting against ISIS in Syria gained great sympathy and support, not only from the Kurdish people but also leftists, socialists, and anarchists in Turkey and around the world. These groups supporting the Kurdish struggle in Syria have been targeted by both radical Islamic factions and the Turkish government. The Suruç massacre (2015), which took place in a Kurdish district in southern Turkey only a few miles from the border town of Kobane in northern Syria, was the first massive attack signifying the spillover of the Syrian war into southern Turkey. The victims consisted of a number of Turkish and Kurdish socialist and anarchist youths who were there as part of their solidarity campaign for the rebuilding of Kobane, which has played an important role for its fighters and civilians in their struggle against ISIS. During the time of my fieldwork, I encountered the neighbours, relatives, and friends of some of the victims of the Suruç massacre. Following the life stories of the victims, I explored new political actors and many conflicts and/or alliances, complicating the existing intricate networks in Gülsuyu-Gülensu neighbourhood. A closer look at the connections provides an alternative perspective for analysing how a transnational development reverberates in the local dynamics of an urban neighbourhood and how neighbourhood politics and international politics beyond national borders are interrelated.

Nazegül Boyraz, a 55-year-old woman with 4 children, had been living in Zümrütevler neighbourhood adjacent to Gülsuyu-Gülensu for many years when she lost her life during the Suruç massacre. Her funeral took place in the İstanbul Gülsuyu Cemevi and Cultural Centre with thousands of people lining the streets of Gülsuyu-Gülensu. Cebrail Günebakan, who was born and grew up in Gülsuyu-Gülensu, was one of the socialists fighting the urban transformation project and drug trafficking in his neighbourhood. The drug gangs shot him on the streets of Gülsuyu-Gülensu on 24 July 2013. He then continued his political activities in different cities and became one of the victims of the massacre. Duygu Tuna, the co-chair of the pro-Kurdish Peoples' Democratic Party (Halkların Demokratik Partisi, HDP) in Maltepe, had also been living in Gülsuyu-Gülensu before she was killed in Suruç. Okan Danacı, who was injured in Suruç, and Büşra Mete, who was killed in Suruç, were friends of Cebrail. They all went to the same high school in the Esentepe neighbourhood adjacent to Gülsuyu-Gülensu and took part in the political activities of the neighbourhood.

These people were active in both local and national politics, and were respected figures in the neighbourhood. Their loss increased concern and fear among the residents and changed local political dynamics. According to one of my interlocutors, who was a member of a radical leftist political organisation during the 1970s and has been living in Gülsuyu-Gülensu for more than 40 years, the loss of young political activists resulted in a 'loss of blood' in the continuous struggle of the neighbourhood:

> The political situation of the neighbourhood is continuously being reversed. We cannot provide collective solutions to our problems or create alternative politics. And when you receive blows one after another, you lose your important cadres. What I mean by cadre is the people, the qualified political cadres. Thirty-three young people went to Suruç. All those were the people who constituted the backbone of that political movement. Then what happened? Thirty-three brains, minds, and experiences were thrown out. Now it takes so much time to get a new thirty-three instead of them. When you think all these together, you can see the recession.

The 'loss of blood' in the neighbourhood is not only limited to the people killed by drug gangs or in the Suruç massacre. These names are the ones who are publicly known because of their loss. As I was told during my fieldwork, many others went to Kobane and/or Rojava to fight for the People's Protection Units (YPG) and/or the Women's Protection Units (YPJ), the armed forces of the Kurdish region of Syria. These cases seem to repeat themselves in the course of the war in Syria, and the loss of young political actors from neighbourhood politics in the battles of a transnational war has emerged as a new reason for the withdrawal from organised politics in Gülsuyu-Gülensu in recent years. On the one hand, the numbers of the politically active members of the neighbourhood decrease because of their imprisonment or murder in the course of both the drug wars and the war in Syria, which leads to a lack of solidarity and resistance organisations. On the other, neighbourhood inhabitants' witnessing of these high prices of political activism creates fatigue around engaging in politics, which is shaped by the affective states of grief, sadness, anxiety, and/or fear.

The loss of neighbourhood youth is not limited to the socialists. Zooming in on the profiles of the gang members also demonstrates the repercussions of the Syrian war in the neighbourhood. During my fieldwork, Ali Şengül was the first but not the last person who brought the impact of the Syrian war to my attention, claiming that the contentious situation in the neighbourhood had changed during the course of the war:

> With the start of the war in Syria, the situation shifted through an ideological line. While the old gang had few people who were selling drugs, now they are an organised gang of almost 100 people. There are different mechanisms like *Büyük Birlik Partisi*, *Alperens* and *Osmanlı Ocakları* behind them. People make *bozkurt* signs now and some of them are Alevi and Kurdish. Taking photos under Ottoman *tughras*, making *bozkurt* signs, certain shape of beard etc. are all signs of them. Thus, I think this process is not only related to a gang or mafia but it is also fuelled by a conservative nationalist ideology.

The above-mentioned far-right organisations not only support the government's conservative nationalist ideology, they are also vigorous advocates of the Turkish state's war against Kurds in Turkey and its interventions in Syria. Furthermore, they either provide support or send militants to Syria to fight Syrian Kurds. The Turkish government's nationalist, pro-Ottoman, and xenophobic discourses against Kurds do not only appeal to Islamist organisations. Besides the Islamists, members of the Nationalist Movement Party (*Milliyetçi Hareket Partisi*, MHP) and the Grey Wolves (*Ülkü Ocakları*, the youth organisation of MHP) have been going to Syria to join ISIS, al-Nusra, or other Islamist groups in the region. Their involvement in the war became publicly known when their funeral ceremonies took place in different parts of the country with the participation of party members and administrators. At the beginning of the military campaign for Afrin on 20 January 2018 in particular, both the government and the nationalists and ultra-nationalists made declarations volunteering to fight the Kurds in Syria.[9]

These developments certainly had a direct impact in Gülsuyu-Gülensu, where the youth of the neighbourhood are broadly divided into two groups: the supporters and the enemies of the Kurds fighting in Syria. While the locals were fearfully following the news, rumours in the neighbourhood made them even more anxious, as one of my interlocutors told me while we were talking about the current political situation in one of the local cafes:

> We heard that some of the gang members went to Syria for the war. It is said that they are fighting with jihadist groups there and even the one who killed that Russian pilot and caused the plane crash stayed here, in our neighbourhood. He stayed at someone's house, can you believe it? That is to say, we are in such a period that these kinds of relationships exist. The problem is not a simple gang. It is about the system that tries to eliminate, to domesticate, or to exterminate this type of neighbourhood. It started with

our neighbourhood. What happened in Gülsuyu was only the beginning and it reached up to the death of a friend. His death was a milestone. Currently, there is not a strong opposition that can fight against such an organised structure.

The gang members' involvement in the state-sponsored violence and their eagerness to fight against Kurds have always been clear to the locals due to their everyday encounters in their neighbourhoods. However, this became a nationwide issue during the hearings into the murder of Hasan Ferit Gedik. On 10 January 2018, during the 35th session of the hearings, Gedik's lawyer showed some photos of one of the accused gang members, those of Uğur Köroğlu, who had been detained after Gedik's murder but later released. The photos show that Köroğlu went to Syria's Türkmen Mountain to fight with jihadists. Then on 31 January 2018, the newspapers shared another report about the gang members' engagement with the Syrian war. Fifteen suspects accused of homicide sent a letter from prison to President Recep Tayyip Erdoğan in which they wrote, 'There are fighters ready to fight in Afrin, waiting for your orders in prisons.'[10] This not only shows a group of gang members' enthusiasm for fighting against Syrian Kurds, but also demonstrates how the Turkish state's nationalist and warmongering discourses have become powerful in local settings. Although the desire to fight against the 'enemies' has not turned into a local conflict in the neighbourhood where the parties of the war – Kurds, Alevis, Islamists, socialists – are living together, tension is always in the air. While Gülsuyu-Gülensu can keep its 'peaceful' neighbourhood image within its borders, it silently witnesses the fight of its 'hostile' neighbours in the battles of the Syrian war.

Karen Büscher (2018), in her article on violent conflicts in Central and East African cities, argues how the nature of socio-economic or political networks within urban neighbourhoods is often strongly connected to conflict dynamics at the national or regional level. Unpacking different layers of increased violence in this particular poor urban neighbourhood gives important insights into both the distinct features of political violence and how it intertwines with (inter)national dynamics. It also has the potential to show how concrete examples of violence connect broader conflicts while retaining distinguishing characteristics shaped by the urban landscapes in which they arise. The politically polarised structure of the neighbourhood with conflicting political powers at the local level also sheds light on the transformation of larger societal dynamics in Turkey. It is remarkable how focusing in at the micro-scale level of the Gülsuyu-Gülensu neighbourhood in Istanbul can reflect the everyday making of international politics, and how international politics can reshape a neighbourhood in return.

Concluding remarks

Drawing on my ethnographic fieldwork in the Gülsuyu-Gülensu neighbourhood, I have argued that instead of discussing the neighbourhood as a homogeneous or heterogeneous entity, looking at the ways in which claims to unity, solidarity, or contention function together allows us to see what politics might mean in such a state of contingency. In that vein, the case study underlines that such political fragmentations not only produce political conflicts among the residents, but also create political potential for collective action. Moreover, the case study shows that these political fragmentations are not always fixed within local dynamics, but are contingent upon changing national/international circumstances. Neighbourhood as a scale makes these developments and the relationships between them more visible – it is an 'awkward scale' (Comaroff and Comaroff 2003) that ties the local and the global together.

Taking neighbourhood as a scale of analysis helps show how urban violence taps into both national and regional conflicts. As most of my interviews demonstrate, the local political conflicts/alliances are not limited to this location. While the neighbourhood is influenced by all kinds of national political developments, it also demonstrates that these internal conflicts/ alliances are the representations of a broader political context. For instance, the resistance to an urban transformation project does not concern actors solely within the neighbourhood, since they have to fight the spatial and ideological threats of the state–capital alliance at the same time. And such a resistance targets not only the entrepreneurs or the state institutions, but also various political actors playing important roles in the struggle for hegemony. Consequently, the ups and downs of this ongoing struggle are not limited to the contentious political actors in the neighbourhood; they are also closely related to both national and international political developments.

Looking at the Gülsuyu-Gülensu case, I posit that a closer look at interwoven local networks of various actors in this neighbourhood has the potential to show the neighbourhood as not only a place of politics but also a dynamic web of relations enabling politics and claim-making by different actors. It also provides insight into how political violence has changed the neighbour- and neighbourhood-making processes and what this means for socio-political transformations in Turkey. The current developments in the neighbourhood in connection to the broader transformations in the region demonstrate that the neighbourhood is simultaneously a space of coexistence, solidarity, hostility, and political contestation that allows certain claims to be made while hindering others.

Notes

1 This chapter is based on ethnographic fieldwork in the framework of the research project 'Political Participation, Emotion and Affect in the Context of Socio-Political Transformations' funded by the German Research Foundation (DFG). The project analyses the transformative potentials and limitations of affective politics for political participation and transformation during and in the aftermaths of the uprisings of 2011 in Egypt and the Gezi uprisings of 2013 in Turkey.

2 This chapter was written while suffering from chronic pain during a long-lasting period of sickness. Thus, all the mistakes and shortcomings belong to me. But I would like to express my gratitude first to the editors Nazan Maksudyan and Hilal Alkan for their critical reading and insightful comments, which contributed to this chapter a lot. I am deeply grateful to Armanc Yıldız for his important contributions and editing. His insights made this chapter richer. Many thanks to my friends Çiçek

İlengiz, Erdem Kayserilioğlu, and Yaara Benger Alaluf for their helpful comments on earlier drafts.

3 *Gecekondu*, with its literary meaning 'landed overnight', refers to the informal housing in the city quarters built with the assumption of them being temporary settlements, but which then became major elements of the big cities, particularly Istanbul and Ankara. Starting from the 1940s, large numbers of peasants moved from rural agricultural areas to urban centres where industry was developing. *Gecekondu* areas have expanded through different waves of immigration and become housing areas of the urban poor. Today, only a few *gecekondu* neighbourhoods remain in Istanbul. Many of them are being transformed and replaced by luxury apartment settlements by the urban transformation projects. For a detailed analysis of the *gecekondu* issue, see (Işık and Pınarcıoğlu 2001; Erman 2011).

4 *Cemevi* is an architectural setting used for the religious and cultural gatherings of Alevis in Turkey.

5 Gedik's funeral was blocked by the police who refused his friends' and family's wishes to have a commemoration ceremony at the site of murder before being buried. His dead body rested in his own neighbourhood's *Cemevi* for three days under the siege of police and water cannons surrounding the neighbourhood. After days of protests and negotiations, Gedik's body was buried in Gazi Neighbourhood Cemetery, another working-class neighbourhood, following commemoration ceremonies and marches with thousands of participants in the Kücük Armutlu, Gülsuyu-Gülensu, and Gazi neighbourhoods. During the protests, commemoration ceremonies, and the funeral, a group of young militants appeared with arms and red masks and took the lead. These scenes attracted great attention from the media and the police immediately launched operations in several locations.

6 https://www.facebook.com/pg/Gizlenen-Ger%C3%A7ekler-447533762025788/photos/?tab=album&album_id=452606151518549 Accessed 17 February 2019.

7 https://www.al-monitor.com/pulse/originals/2013/10/istanbul-gulsuyu-crime-neglect.html Accessed 12 February 2019.

8 Based on the numbers from November–December 2015, 1,621 registered Syrians were living in Maltepe, where the Gülsuyu-Gülensu neighbourhood is located. Although the local governments have been significant actors working directly with refugees, neither Maltepe District Governorship nor Maltepe Municipality has reliable records of refugees or systematic research on the issue in their district. They only implemented a project in association with the United Nations High Commissioner for Refugees (UNHCR) entitled 'The War Is Over, Now Is Time to Recover Mental Health of Syrian Refugee Children in Maltepe' between 1 August 2017 and 28 February 2018. As they shared the results of the project on their website, the project reached 354 Syrian children in 183 households. For details of the project, see http://projemaltepe.gov.tr/suriyeli-cocuklar-projesi-tamamlandi/.

9 http://www.hurriyetdailynews.com/nationalist-leader-bahceli-says-he-is-ready-to-sacrifice-his-life-in-turkeys-afrin-operation-126865 Accessed 12 March 2019.

10 https://m.bianet.org/bianet/insan-haklari/193,856-hasan-ferit-gedik-in-katil-saniklarindan-cumhurbaskanina-mektup-afrin-de-savasmaya-haziriz Accessed 12 March 2019.

References

Aslan, Şükrü. 2004. *1 Mayıs Mahallesi: 1980 öncesi toplumsal mücadeleler ve kent [Mayday Neighbourhood: Social Struggles and the City before 1980]*. İstanbul: İletişim.

Aslan, Şükrü. 2010. 'Bir Politik Mekan Olarak "Gül" ve "Su"yun Mahallinde Toplumsal ve Kültürel Örüntülere Tarihsel Yolculuk.' Accessed 13 February 2018. www.arkitera.com/

gorus/bir-politik-mekan-olarak-gul-ve-suyun-mahallinde-toplumsal-ve-kulturel-oruntulere-tarihsel-yolculuk/.

Bartu Candan, Ayfer and Cenk Özbay. 2014. *Yeni İstanbul Çalışmaları: Sınırlar, Mücadeleler Açılımlar*. İstanbul: Metis.

Brenner, Neil. 2004. *New State Spaces: Urban Governance and the Rescaling of Statehood*. Oxford: Oxford University Press.

Büscher, Karen. 2018. 'African Cities and Violent Conflict: The Urban Dimension of Conflict and Post Conflict Dynamics in Central and Eastern Africa.' *Journal of Eastern African Studies* 12 (2): 193–210.

Çavuşoğlu, Erbatur and Julia Strutz. 2014. 'We'll Come and Demolish Your House! The Role of Spatial (Re)Production in the Neoliberal Hegemonic Politics of Turkey.' In *Turkey Reframed*, edited by İsmet Akça, Ahmet Bekmen, and Begüm Özden Fırat, 141–153. London: Pluto Press.

Comaroff, Jean and John L. Comaroff. 2003. 'Ethnography on an Awkward Scale: Postcolonial Anthropology and the Violence of Abstraction.' *Ethnography* 4 (2): 147–179.

Comaroff, Jean and John L. Comaroff. 2016. *The Truth about Crime: Sovereignty, Knowledge, Social Order*. Chicago, IL: University of Chicago Press.

Cörüt, Ilker and Gönül Evren. 2007. 'From Almus to Küçükarmutlu: An Ethnographic Study Tour of the Rural and Sub-urban Space in Relation to State and Market Intrusions.' *Journal of Historical Studies* 5: 33–67.

Ergin, Nezihe Başak and Helga Rittersberger Tılıç. 2013. 'The Right to the City: Right(s) to "Possible-Impossible" Versus a Mere Slogan in Practice?' In *Understanding the City: Henri Lefebvre and Urban Studies*, edited by Gülçin Erdi-Lelandais, 37–68. Newcastle: Cambridge Scholars Publishing.

Erman, Tahire. 2011. 'Understanding the Experiences of the Politics of Urbanization in Two *gecekondu* (Squatter) Neighbourhoods: Ethnography in the Urban Periphery of Ankara, Turkey.' *Urban Anthropology* 40 (1/2): 67–108.

Glebbeek, Marie-Louise and Kees Koonings. 2015. 'Between Morro and Asfalto: Violence, Insecurity and Socio-spatial Segregation in Latin American Cities.' *Habitat International* 54 (1): 3–9.

Graham, Stephen. 2011. *Cities Under Siege: The New Military Urbanism*. London: Verso Books.

Harvey, David. 2003. *The New Imperialism*. Oxford: Oxford University Press.

Işık, Oğuz and Melih Pınarcıoğlu. 2001. *Nöbetleşe yoksulluk: gecekondulaşma ve kent yoksulları Sultanbeyli örneği [Poverty in Turns: Gecekonduization and Urban Poor, the Case of Sultanbeyli]*. Istanbul: İletişim.

Karandinos, George, Laurie Hart, Fernando M. Castrillo, and Philippe Bourgois. 2014. 'The Moral Economy of Violence in the US Inner City.' *Current Anthropology* 55 (1): 1–22.

Kolluoğlu, Biray. 2014. 'Sunuş: Şehre Gören Gözlerle Bakmak.' In *Yeni İstanbul Çalışmaları: Sınırlar, Mücadeleler Açılımlar*, edited by Ayfer B. Candan and Cenk Özbay, 19–25. İstanbul: Metis.

Koonings, Kees and Dirk Kruijt. 2015. *Violence and Resilience in Latin American Cities*. London: Zed Books.

Kuyucu, Tuna, and Özlem Ünsal. 2010. '"Urban Transformation" as State-led Property Transfer: An Analysis of Two Cases of Urban Renewal in Istanbul.' *Urban Studies* 47 (7): 1479–1499.

Lefebvre, Henri. 1991. *The Production of Space*. Oxford: Blackwell Publishing.

Lovering, John and Hade Türkmen. 2011. 'Bulldozer Neo-liberalism in Istanbul: The State-led Construction of Property Markets, and the Displacement of the Urban Poor.' *International Planning Studies* 16 (1): 73–96.

Massey, Doreen. 1994. *Space, Place, and Gender*. Minneapolis: University of Minnesota Press.

Massey, Doreen. 1999. 'On Space and the City.' In *City Worlds*, edited by Doreen Massey, John Allen, and Steve Pile, 151–174. London: Routledge.

Özdemir, Esin. 2013. 'Contesting Neoliberal Urbanisation: Contemporary Urban Movements in Istanbul, The Case of Gülsuyu-Gülensu Neighbourhoods.' Unpublished PhD Thesis. METU, Ankara.

Özdemir, Esin and Ayda Eraydın. 2017. 'Fragmentation in Urban Movements: The Role of Urban Planning Processes.' *International Journal of Urban and Regional Research* 41 (5): 727–748.

Parenti, Christian. 2000. 'Crime as Social Control.' *Social Justice* 27 (3): 43–49.

Peck, Jamie. 2002. 'Political Economies of Scale: Fast Policy, Interscalar Relations and Neoliberal Workfare.' *Economic Geography* 78 (3): 332–360.

Şen, Besime. 2010. 'Gülsuyu-Gülensu ve Başıbüyük deneyimleri [The Experiences of Gülsuyu-Gülensu and Başıbüyük].' In *Tarih, sınıflar ve kent [History, Classes and the City]*, edited by Besime Şen and A. Ekber Doğan, 309–353. Ankara: Dipnot Yayınları.

Şen, Besime. 2013. *The Production of Autonomous Settlements for the Working Class by the Turkish Socialist Movements*. Unpublished presentation paper.

Smith, Neil. 1993. 'Homeless/Global: Scaling Places.' In *Mapping the Futures: Local Cultures, Global Change*, edited by Jon Bird, Barry Curtis, Tim Putnam, George Robertson, and Lisa Tickner, 87–119. New York: Routledge.

Swyngedouw, Erik. 1997. 'Neither Global nor Local: "Glocalization" and the Politics of Scale.' In *Spaces of Globalization*, edited by Kevin R. Cox, 137–166. New York: Guilfor.

Swyngedouw, Erik. 1998. 'Homing In and Spacing Out: Re-configuring Scale.' In *Europa im Globalisieringsprozess von Wirtschaft und Gesellschaft*, edited by H. Gebhardt, G. Heinrity, and R. Weissner, 81–100. Stuttgart: Franz Steiner Verlag.

Tilly, Charles and Sidney Tarrow. 2007. *Contentious Politics*. Boulder, CO: Paradigm Publishers.

Wacquant, Loic. 1999. 'Urban Marginality in the Coming Millennium.' *Urban Studies* 36 (10): 1639–1647.

Yalçıntan, Murat C. and Erbatur Çavuşoğlu. 2013. 'Kentsel Dönüşümü ve Kentsel Muhalefeti Kent Hakkı Üzerinden Düşünmek.' In *Kentsel Dönüşüm ve İnsan Hakları*, edited by Turgut Tarhanlı, 87–106. İstanbul: İstanbul Bilgi Üniversitesi Yayınları.

Yıldız, Erdoğan. 2013. *Kendi Sesinden Gülsuyu-Gülensu [Gülsuyu-Gülensu from Its Own Voice]*. Ankara: Nota Bene Yayınları.

Yonucu, Deniz. 2018. 'Urban Vigilantism: A Study of Anti-Terror Law, Politics and Policing in Istanbul.' *International Journal of Urban and Regional Research* 2 (3): 408–422.

11 Urban tectonics and lifestyles in motion

Affective and spatial negotiations of belonging in Tophane, Istanbul

Urszula Ewa Woźniak

Stepping out of the neighbourhood house (*semt konağı*), the neighbourhood representative (*muhtar*) and I begin to walk towards her office. The weekly session of the people's parliament (*Halk Meclisi*) had just adjourned. This was not a regular session, just as this was not a regular week. The city was witnessing day four of the proclaimed 'democracy festival', following what is now known to have been the night of a military coup attempt.

Ordinarily, one or more local *muhtar* and a member of the municipal council host the session at the neighbourhood house. While a *muhtar* was present at this session, the member of the council only arrived late, just half an hour prior to its closing. Both of their phones kept ringing. The daily business on the agenda had been put aside and the time instead used for sharing personal stories of how the night was experienced. Several people present, mostly male, walked up to the *muhtar*'s desk, phones in hand, to share the pictures they presumably took during the night.

Now, as the *muhtar* and I step out of the neighbourhood house, we continue to be approached by people with their alleged stories of martyrdom. As we keep walking, a man, whom I estimate to be in his late 30s, comes running up to us. He whips out his phone to show some photographs to the *muhtar*. While I do not get a glimpse of them, I hear him calling out: 'My *muhtar*, we were all out in Taksim that night. All of Tophane was there, beating up all of Cihangir!'[1]

While the night of the 15 July coup attempt in Turkey has since triggered many inquiries into its wider socio-political implications in terms of authoritarianism and societal polarisation, little attention has been paid to the embeddedness of the event in urban space in Istanbul, and negotiations of political and affective belonging. This chapter inquires into the times and specific places in which the formation of affective ties to specific neighbourhoods – in both extraordinary and everyday times – becomes evident in the form of a clash over urban territories.

Although both the Gezi uprising of 2013 and the coup attempt in 2016 reinforced the already-growing social and spatial polarisation, urban space in Istanbul, and in particular the neighbourhood (*mahalle*), has long been the subject of political and ideological contestation. As I show in this chapter, the trope of a clash of neighbourhoods actually demonstrates a territorialised conflict over lifestyle. Over the past two decades, this conflict has been particularly apparent in the larger district of Beyoğlu, which includes the neighbourhood of Tophane. Negotiations around the notion of being *mahalleli* – the one of/from the *mahalle* – are more than a reflection of constructions of residential and affective belonging. Looking into practices and narratives of changing landscapes of consumption in the neighbourhood of Tophane, I argue that the governing Justice and Development Party's (*Adalet ve Kalkınma Partisi*, AKP) public discourse on the moral fabric of certain neighbourhoods feeds into everyday contestations around belonging to specific urban *mahalle* spaces. Analysing negotiations of *mahalleli*-ness as a form of urban kin makes it possible to scrutinise moral order and urban relatedness between continuity and ambivalence. The following inquiry unfolds the multiple ways in which mundane conflicts in the neighbourhood are linked to macro-level political and socio-economic tensions.

Drawing boundaries of inclusion and exclusion: the *mahalleli* as urban kin

As an emic term, the notion of *mahalleli* frequently appeared across varying social and local settings throughout my fieldwork. In the following, I argue that the *mahalleli* notion reflects the construction of an urban kin that renders the *mahalle* community itself 'a new and agonistic territory for the organisation of political and ethical conflicts' (Rose 1996, 337).

The urban kin of the *mahalleli* includes a sense of moral obligation, solidarity, and mutual assistance between those people it creates as members of the neighbourhood community. The earliest conceptualisations of these kinship-like relations closely tied them to the idea of family: '[T]he concept of neighbourhood [...] pinpoints a field of social relations in which local association is suffused with the ideology and values of kinship and is thus drawn into the familial domain' (Fortes 1969, 245). Forms of urban kinship are therefore hierarchical, reflecting gendered and patriarchal societal configurations. Just as with any form of kinship ties, the social ties of *mahalleli*-ness are reproduced as rooted in descent and thus as an inevitable, natural state. However, in practice, the reproduction of social and spatial relatedness through these ties actually involves active work (Bjarnesen and Utas 2018, 5).

The construction of the *mahalleli* as urban kin is thus related to two levels of affective and political belonging. First, reproducing ideas about how to live together properly in solidarity and with mutual obligations is often related to a constructed 'other'. Thus, in their narrations, my interlocutors often referred to the *mahalleli* as a figure that potentially represents, safeguards, and defends a given neighbourhood against outside threats to its moral order. Second, conflicts around belonging to a specific *mahalle* reflect a moralised duality of opposing positions that mirror claims to political authority and legitimacy.

The social web of relations of the *mahalle* varies in intensity and intimacy, ranging from more stable networks, to durable engagements, to more coincidental and fluid forms of encounter. While permanent residency is thus not necessary to be considered a *mahalleli*, it does require some kind of permanence in physical presence in the neighbourhood.[2] The very recognisability of people and expectation of their presence in the neighbourhood lie at the heart of the historically normative definition of the *mahalle* in the sense of 'public familiarity' (Blokland 2017, 86ff.); being seen and known by other people in the *mahalle* determines social relatedness. It furthermore creates a feeling of safety, which is what makes it attractive for governing bodies to capitalise upon.

As shown elsewhere (Woźniak 2018a, 78f.), the notion of the *mahalleli* is often juxtaposed with the category of the 'guest' and the 'stranger' in a given neighbourhood. The differentiation between the stranger as 'other' and the neighbour, as a process of negative distancing, serves to conceal various forms of social difference and the fact 'that some-bodies are already recognized as stranger' (Ahmed 2000, 3). While this first and foremost alludes to the visibility of markers of difference such as race and gender, it also applies to other markers and lifestyle practices discussed in this chapter. Finally, as shown by the example of Tophane, the boundary-drawing of who belongs to the *mahalle* as *mahalleli* and who does not also reflects both the scale and effects of processes of gentrification and urban transformation policies on the social make-up of the neighbourhoods. In the following sections, I show how various forms of social anxiety underpin these processes of boundary-making.

Modernity, lifestyle clashes, and nostalgia: class encounters in the *mahalle*

Historically, the construction of social relatedness in and through the *mahalle* relied on the socio-spatial proximity of its residents, which allowed for intimate social ties and micro-level support networks (Behar 2003). Over the past 17 years, the AKP's inherently spatial efforts to establish and maintain hegemony (Çavuşoğlu and Strutz 2014) through a huge number of oftentimes large-scale urban transformation projects have threatened these ties. Thus, many of Istanbul's historic *mahalle* spaces have become subject to new forms of gentrification and commodification, attracting a new middle class seeking to live in inner-city locations with its own set of lifestyle and consumption patterns (Soytemel and Şen 2014). Despite the long-standing efforts of anthropologists and sociologists of Turkey to complicate the urban encounter between middle-class and lower-class urbanites, many explanatory models tend to perpetuate the established lifestyle dualism of secular urban middle classes versus pious rural migrants, thereby reproducing a duality of modern versus anti-modern neighbourhoods.

Lifestyle debates in relation to urban territoriality have long formed the core of discussions on modernity. Several scholars of Western and non-Western urbanisms alike have pointed to the multiplicity and pluralism of urban modernity, and have argued for disentangling and de-territorialising the relation between the city and modernity by taking multiple times and geographies into account (Dibazar et al. 2013; Robinson 2013) and decoupling modernity from its association with the (urban) West (Robinson 2013; Türeli 2018, 7). Postcolonial scholars in particular have shown that, in academic and popular discourses alike, modernity has served as a spatio-temporal bias, dividing the world – including urban territories – into various hierarchised binaries such as the modern and developing (Dibazar et al. 2013, 676; Robinson 2006, 7), of being either ahead of or behind time (Ahıska 2003, 354). Various attempts at spatial and temporal diversification of the modern have been undertaken with the help of concepts such as 'hybrid modernities' and 'alternative modernities' (Gaonkar 1999). Criticising this revision of the modern, Gurminder Bhambra argued to view 'the very concept of modernity itself as problematic' (2007, 2), a critique which has also been employed against the modernisation project in Turkey, be it the state-led modernisation project of the 1930s or the market-inspired modernisation of the 1990s (Özyürek 2006, 18).

Despite this theoretical critique of linear understandings of urban modernity in the Turkish context, the effort to detach the notion of urban modernity from particular territories fixed in certain times has not yet come to fruition. Thus, urban scholars who focus on Turkey have long been especially preoccupied with the question of integrating migrants from Anatolia living in the so-called *gecekondu*[3] squatter settlements as part of a 'distorted urbanisation […] of peasants who had failed to urbanise themselves' (Aslan and Erman 2014, 95). Some authors have discussed the various modes of social distinction, including forms of satire and humour, which continuously paint a picture of rural outsiders in need of domestication (Öncü 2002). Others have underlined that these categories of rural outsider versus urban establishment are reproduced by the actors involved, such as through the self-monitoring of behaviour along these established lines (Özyeğin 2002, 65). On the one hand, the ambiguity of the class identity of the newcomers and their lifestyles is discerned to have triggered anxiety among the former middle classes from those areas (Örs 2018, 83). On the other hand, Amy Mills' analysis of narratives on the Kuzguncuk neighbourhood's past multi-ethnic cosmopolitanism underlines that the urban middle classes' attraction to (formerly) multi-ethnic areas (Jackson and Butler 2015; Mills 2010, 2018; Örs 2018) helps to gloss over current divisions of class and origin (2006, 363).

Notwithstanding the fact that encounters across class and ethnicity have been shaping Istanbul's urbanism for decades, scholarly attention has mainly been on the supposed anxieties of the established urban middle class and its needs for social and cultural distinction. This ignores evidence that both the established urbanite and its constructed 'other' are culturally transformed by their encounter, as are their lifestyle practices (Türeli 2018, 5). Instead, it reproduces the idea of rural lifestyles that are seemingly antithetical to the aspired-to Western-style urban modernity. Anxieties towards a perceived 'other' – though for different reasons – also preoccupy the lower classes. The example of the district of Beyoğlu and its Tophane neighbourhood discussed in this chapter moves beyond the dualisms inherent to these discussions.

A neighbourhood of conflictual encounters: Tophane

My analysis of the historical neighbourhood of Tophane presented in this section is part of a larger comparative study within the scope of my PhD thesis. In it, I study the two neighbourhoods of Tophane and Kurtuluş to demystify the alleged monolithic spatiality and naturalised history of the *mahalle* by instead showing the relationality of both neighbourhood spaces.

While having drawn scant attention from urban scholars, Tophane's past and present are shaped by multiple layers of diversity. Having been a settlement of Christian minorities in Ottoman times, the ongoing Turkification since the foundation of the republic as well as the violent attacks on non-Muslims in 1955 known as the 'September 6–7 Events' led to their violent expulsion from the neighbourhood. Nevertheless, during my research, narrations by several older residents of the neighbourhood attested to the presence of Greek and other minorities in Tophane and its surroundings up until the 1970s. Some interviewees even made historical references to the Armenian porters of the neighbourhood from the late Ottoman and early Republican eras when describing the diverse heritage of the neighbourhood. From the 1950s onwards, Istanbul's urbanisation brought a new kind of diversity to the neighbourhood, namely migrants from the towns of Siirt and Bitlis as well as some from the Diyarbakır, Van and Bingöl provinces. These groups continue to constitute a large portion of the present neighbourhood population. Today, Tophane is far from being homogenous and is in fact polyglot: Next to many Kurds, who mostly settled there in the 1990s, many families from Siirt still speak Arabic, as do Syrian refugees who came to the neighbourhood during the Syrian war.

Over the past two decades, multiple processes of urban transformation[4] as well as its proximity to the two economically more advantaged neighbourhoods of Galata and Cihangir have significantly impacted the demographic make-up of Tophane. Rising real estate values in Tophane reflect the impact of the inflow of high-income residents and shopkeepers as well as speculative investment in the neighbourhood.[5] Although the main artery of the neighbourhood, Boğazkesen Street, has been almost completely commodified over the past few years, attracting art galleries, high-level restaurants, hotels, and shops, many of these spaces are currently empty, looking for new owners following the political and economic turmoil of the period following the coup attempt of 15 July.

These processes have threatened both the former socio-economic and cultural and moral orders of the neighbourhood. Tophane is known as a stronghold of AKP voters, with a large population of pious residents, while simultaneously being in close spatial proximity to various nightlife hubs. Over the past decade, there have been several vigilante incidents related to alcohol consumption in the public and semi-public spaces of the neighbourhood, both during the month of Ramadan and on other occasions (Massicard 2018; Öz and Eder 2018; Woźniak

2018b).[6] The targets of these assaults were of middle-class background, many of them working in the creative sector.

The wide coverage of these incidents by both pro-government and oppositional media outlets has contributed to the public image of the collective will of the neighbourhood community as well as the pious, Muslim *mahalleli* of Tophane (Başaran 2015; Tophanehaber 2016). This mediated idea of the *Tophaneli* (meaning 'a person from Tophane') has also relied on the construction of the neighbourhood's 'other'. In the framework of this discourse, the adjacent neighbourhood of Cihangir serves a narrative of two neighbourhood populations whose lifestyle choices, educational backgrounds, and socio-economic profiles seemingly force them to co-exist in 'dangerous proximity' (Öz and Eder 2018).

Urban tectonics in motion

Let us return to the opening vignette of this chapter. It portrays a very specific experience, usage, and construction affectively, as well as politically, of one *mahalle*, Tophane, in direct enmity with another, Cihangir, at a particular moment in time.

Though the session in the *Semt Konağı* a few days after the 15 July coup attempt included highly emotive displays of alliance with the counter-coup fighters and celebrations of martyrdom and self-defence, observing this encounter between the *muhtar* and a self-identified *mahalleli* of Tophane went beyond the scope of the mere coup attempt. While the portrayal of heroism and bravery as a sure fact was certainly questionable, its spatial dimension instantly intrigued me. It portrayed a *mahalleli* of Tophane as an unquestionably heroic figure, who would readily take up the state's call to self-defence and the protection of democracy that night by seemingly fearlessly swarming to Taksim Square. Moreover, the narration painted a picture of a supposedly monolithic, unified, and indivisible body of the *mahalle* moving towards Taksim Square in the middle of the night, fuelled by the idea of not only defending the square, but both symbolically and physically fighting another *mahalle*. This picture of the clash of two *mahalles* constructs each as a pure and homogenous space in the face of the 'perception of danger posed by outsiders to moral and social health or well-being' (Ahmed 2000, 26). Here, the danger is constructed as an inherent threat to national unity. The brutality of this image thus lies in the equation of a pro-government stance with one *mahalle*, and a putschist attitude with the other, with the only legitimate act being violent aggression.

The self-identified *mahalleli* who ran up to the neighbourhood representative in the aftermath of the coup attempt was making sense of the public enemy by evoking the idea of violently attacking the bordering neighbourhood of Cihangir. This conflation of two discourses – that of an enemy of the state and an enemy of the *mahalle* – is one of an alleged neighbourhood war between Tophane and Cihangir. This reiterates the idea that the construction of the urban kin of the *mahalleli* relies on a negative definition of its 'other'.

In the past two decades, urban scholars have used the metaphor of 'social tectonics' to describe the minimal contact between differently classed and raced groups in newly gentrified neighbourhoods (Butler and Robson 2001; Jackson and Butler 2015). With the help of this image they emphasise that these groups move past each other like tectonic plates with little contact (ibid., 2355f.). My understanding of urban tectonics reaches beyond the idea that there is limited (or only friction-prone) contact between old and new users of gentrified neighbourhoods. Rather, this chapter questions the seemingly inescapable character of these sometimes earthquake-like collisions by contextualising them both historically and politically. Taking the public discourse surrounding the two neighbourhoods at hand into account allows for a deeper understanding of the political value of these collisions. The following section thus analyses the discursively reiterated tension between Tophane and Cihangir against the backdrop of the ideological disputes over the district of Beyoğlu in both the past and present.

Beyoğlu: home of the (not yet) modern and marginal

Both the neighbourhoods of Tophane and Cihangir belong to the district of Beyoğlu, which has been governed by the AKP since 2001. With economic neo-liberalisation unfolding from the beginning of the 1980s, and rising anticipation of Istanbul's rebirth as a global city (Keyder 1999), the cultural heritage of Beyoğlu has been increasingly subject to contestation.

Following the cleansing, rehabilitation, and demolition of the district's more deprived neighbourhoods such as Tarlabaşı beginning in the 1980s, normative contestations of what Beyoğlu culturally represented intensified. These alternated between the nostalgic accounts of an Ottoman Christian and European elite residing in Pera[7] from the 18th century onwards, on the one hand, to condemnations of the moral decay of the district as the 'brothel' of the city on the other. In the 1990s, the district continued to be rendered a site of ideological struggle, especially in the aftermath of the first ground-breaking electoral success of the Welfare Party (*Refah Partisi*, RP) in the municipal elections of 1994, which included a victory for its candidates to the Istanbul Metropolitan Municipality. With Istanbul being – for the first time in the history of the republic – governed by an Islamist party, Beyoğlu became the core site of ideological struggle between secular and Islamist forces and with it, the constructed duality of lifestyle clashes was once more on the agenda of urban scholars and other academics analysing Istanbul (Türeli 2018, 9). Given a new urban visibility of Islamic lifestyles in Istanbul and the, partly highly anxious, perception of this political event as a 'reconquest of the city' by Islamists (Bartu 1999, 2001), interest in the cultural fate of Beyoğlu, a hub of the city's entertainment industry, was rising.

Fast-forward more than two decades and the trope of Beyoğlu as the territory of contested lifestyles lives on in President Erdoğan's speeches. His own ties to Beyoğlu are manifold. A child of the district's Kasımpaşa neighbourhood, he first became the RP's Beyoğlu district chair in 1984 and was elected mayor of Istanbul in the 1994 surprise victory a decade later. In March 2018, Erdoğan contended that Beyoğlu was housing the 'marginal' members of society.

> As long as these marginals that we see appearing on the streets of Beyoğlu behave in a civilised manner, they can remain to be one of the colours of this country. But if they go as far as to show off pressure, intolerance, aggression, and violence against those who are not like them, with all due respect, we will pull them by their ears and throw them to the place where they belong. [...] No one has the right to disrupt the peace of this country, to attack the values of this nation.
>
> (Erdoğan: Beyoğlu'ndaki marjinaller, 23 March 2018)

Here, Erdoğan designates the politically, but also culturally, marginal as disturbers of the peace, and locates them and their seemingly divergent lifestyles as belonging to the 'streets of Beyoğlu'. While at first glance, Erdoğan appears to recognise the pluralism of lifestyles ('the colours of this country'), he immediately subordinates the Beyoğlu lifestyle. He places the district, famous for its music and clubbing scene, but (formerly) also for its prostitution and gay and queer scene, at the lowest rank of the moral hierarchy. In short, Beyoğlu seems to stand for a hedonistic and impious lifestyle.

The term 'marginal' rose to prominence in the political discourse in Turkey from the 1980s onwards. Denoting various political and cultural 'others', including anti-communists, queer and non-hetero-normative sexuality and lifestyles (Bora 2018, 67f.), it derives its discursive power from excluding the groups it is wielded against. In this statement, Erdoğan furthermore makes use of his very frequent formulation 'with due respect' (*'kimse kusura bakmasın'*) (ibid., 108–116), leaving no room for interpretation about the fact that a vigilante action, 'pulling them [i.e. the marginals] by their ears', is appropriate.

A closer look at the statement's context reveals in more detail who the abovementioned marginals are: 'We will not be tolerant towards the marginal organisations who keep some of our universities from doing their actual jobs and allow certain groups to turn the universities into protest sites' (Erdoğan: Beyoğlu'ndaki marjinaller, 23 March 2018). His statement came as a reaction to a violent confrontation between two student groups at Bosphorus University. On 18 March 2018, a group of students celebrated the incursion of Turkish military forces into the Syrian town of Afrin by distributing Turkish delight on campus, to which another group of students quickly reacted with an anti-war counter-protest. The police intervention that followed the escalation included the detention of protesters critical of the pro-military group for weeks after the actual confrontation. In his lengthy statement, Erdoğan leaves no doubt that he views the protesting students as looters (*çapulcular*), first coined as a derogatory term for depicting the protesters by then Prime Minister Erdoğan during the Gezi uprising of 2013. This rhetorical positioning of anti-government student protesters with the residents of Beyoğlu and the Gezi protesters shows the urgency of analysing political mobilisation of urban tectonics within the complex landscapes of lifestyle in more detail. As the ethnographic analysis below shows, the discourse of territorialised lifestyles reaches beyond notions of the modern, tying it to questions of political legitimacy and national belonging.

Changing landscapes of lifestyle and consumption

Historical *mahalle* spaces such as Tophane have increasingly become sites of new consumption spaces, which indicate the changing class and taste boundaries running through the neighbourhood. These new places and practices of consumption both manifest overt differences in lifestyle, and reflect the changing eating and drinking habits that continue to be subject to ideological contestation. This holds especially true for the visible consumption of alcohol in the residential areas of the *mahalle*. While high-class Islamic café house culture is increasingly entering various neighbourhoods across the city, Tophane has been exempt from this trend, likely also due to it bordering tourist areas such as Galata and Karaköy, in which alcohol is part of the infrastructure serving tourists from around the globe.

In this section, I consider the various functions of different places of consumption. The distinct levels of income they are catering to do not determine the moral, social, and political qualities attached to them. As Lara Deeb and Mona Harb have convincingly shown, what makes a café morally acceptable is often contradictory, rendering urban modernity, once more, multiple and contested (2013a, 2013b). As my analysis shows, practices and places of consumption are also an expression of neighbourly relations, and thus formations of *mahalleli*-ness.

The 'Tophaneli'

Walking by the liquor store located close to the waterfront area of Tophane, I do not fully trust my eyes. There it is, spelled out in bold italic letters, '*Tophaneli*'. The *mahalleli* of Tophane graces the sign of this store, the front windows of which mainly display alcohol, cigars, and cigarettes. For the time being, I am convinced that the owners of this new-looking store must be poking fun at the stereotype of the bigoted, pious, non-drinking *Tophaneli* constructed by several media outlets.

Asking the storeowners for an interview a few days later (13 December 2016), I am quickly proven wrong. I meet with Alpaslan[8] and his father Bahadır, and speak to them several times over the course of the next few weeks. Having started to rent the shop a few years after the coup d'état in 1980, the father has been renting the space for more than three decades now. Bahadır's family originally migrated from Siirt to Tophane, and he was born and raised in the neighbourhood. For many years, Bahadır worked as a primary school teacher in the adjacent Cihangir neighbourhood, where he continues to live to this day. Father and son tell me that this part of the neighbourhood, while being a fully commercialised 'tourist' area now, used to be a residential area up until the 1980s, with village-like qualities, including a chicken slaughterhouse. Their narrations reflect the swift social changes that occurred in the neighbourhood. While the docks served commercial import-export activities, mostly with Russia, up until the 1990s, by the early 2000s they had begun to welcome international cruise ships and their passengers.

Contrary to the reputation of Tophane being predominantly pious, both father and son speak of their shop being located in the historical harbour area. According to them, drinking was part of the local neighbourhood culture in 1970s and 1980s Tophane. While Islamic associations were always present in Tophane, to their knowledge, the numbers of active members in the neighbourhood have increased since the AKP has been in power. Alpaslan and Bahadır navigate the moral ambivalences connected to their economic engagement with alcohol and cigarettes. Adhering to the Muslim faith themselves, they continue to sell alcohol during Ramadan, even though the father explains that he 'morally' would prefer not to. Referring to the existence of merely one more liquor store located in the uphill parts of Tophane, they emphasise the longevity of the presence of both of these stores in the neighbourhood being the factor that allows it. Given that their liquor store is near a mosque, to the regret of many of their tourist customers, it is morally compromising as well as legally prohibited to consume alcohol in front of their store.

Even though many pious fellow *mahalleli* do not approve of their profession, due to the length of their presence in the neighbourhood, father and son are still met with respect. Whether rooted in biographical facts or narratively constructed, long-term residence is a way to legitimise belonging to a *mahalle*. During my fieldwork, many of those who considered themselves *mahalleli* claimed to have resided in Tophane 'for more than 100 years', when their family's migration trajectories would often prove these narratives wrong. These exclamations actively reproduce the idea of the urban kin of the *mahalleli* as an inevitable social tie rooted in ancestral ties to the neighbourhood.

The shop bears Bahadır's nickname, they proudly explain to me. They refer to the father's 'fast' lifestyle as a 'Tophaneli', and while they don't reveal details, they imply a readiness to use violence during his youth being what earned him this nickname. Even though their biographies and occupations belie the idea of the uniformly pious *Tophaneli* portrayed by various media outlets in the aftermath of the vigilante incidents in the neighbourhood, they insist on the interconnection of the *Tophaneli* and the *kabadayı*.[9] According to them, the latter represents a certain degree of danger, as it is closely connected to violence. What does the use of this 'archetype' figure reveal?

The persistence of the *kabadayı*

There is virtually neither a contemporary nor a historic study on Tophane that omits the *kabadayı*, a figuration from the Ottoman past (Başaran 2015; Hür 2010; Lévy-Aksu 2013; Massicard 2018). Ayşe Hür (2010) has suggested nothing less than the defence of the 'honour of the neighbourhood' (*mahallenin namusu*) as being the main duty of the former *kabadayı*. In her contribution on the forms and meanings of vigilante violence in present-day Tophane, Élise Massicard points to the difference between the historical figures of the *kabadayı* and its contemporary form. While the *kabadayı* were detectable individuals, subject to heroisation by their residential communities, the figure has today become a cipher for vigilante action (2018, 14). Similarly, during my fieldwork, the *kabadayı* was frequently referred to anonymously and oftentimes in plural form. Thus, when the *kabadayı* strikes, he does so unidentifiably, which allows responsibility for the often extremely violent attacks to be assigned to an anonymous collective.

This was also reflected in my interview with Ramazan, founder of and journalist for *Tophanehaber* and member of the local Muslim association 'Gönülbağı Vakfı', which has been active in the neighbourhood for about a decade. Expressing his inclination towards this figure, he made it the logo of both *Tophanehaber* and the association Gönülbağı (Figure 11.1).

Ramazan insists that the *kabadayı* first and foremost acts to protect the families of the *mahalle* and their core values. The family as an abstract value

Figure 11.1 Logo of Gönülbağı Vakfı.
Source: Photo by Urszula Ewa Woźniak, 2019.

was central to many narrations justifying vigilante action in Tophane. During my fieldwork, I came across many stories of sexual relations considered inappropriate, of loud sexual interactions behind open curtains – in short, 'immoral' actions visible and audible to the semi-public of the *mahalle*. While these narrations mentioned neither the names of alleged perpetrators nor of the injured families, the vigilante incidents were said to be incompatible with the 'traditional' family structures of the *mahalle*. Further, they were often framed as stories of moral decay in a presumably once morally upright neighbourhood allegedly rid of its former drug problem with the help of fearless *kabadayı*.[10] Once more, relatedness in the city builds here on ideas of 'enduring socio-cultural legacies, explicitly [...] articulated in the language of [– or here, in the name of] protecting – family' (Bjarnesen and Utas 2018, 5).

Ramazan continues to speak about the fighting culture (*kavga kültürü*) of Tophane, which according to him, still prevails in the neighbourhood, 'even if to a lesser extent than earlier' (interview, 20 July 2016). This term was often used interchangeably with the broader 'neighbourhood culture' (*mahalle kültürü*) by several of my interlocutors in Tophane. Sometimes described as a gendered form of courtesy, such as the duty to assist a woman with her heavy shopping bags, it was also described as a defensive culture in some ways. For Ebru, a 32-year-old former gallery employee born and raised in Tophane, what is dubbed 'neighbourhood culture' is the resistance to the massive demographic and class shifts in the neighbourhood. Acknowledging the fact that many former residents of Tophane have sold their places and moved elsewhere over the past decade, ever since the first announcement of the nearby Galataport redevelopment project attracted new investors into the neighbourhood, Ebru speaks of a process of 'othering' that *Tophaneli* resist, 'The attacks against alcohol and on the arts are actually attacks against the forms of "othering" that the *mahalleli* experience' (interview, 19 July 2016).

Locating the 'marginals'

While encounters with and thereby the making of the 'other' or stranger to the *mahalleli* are by far not always violent in Tophane, they often evolve around a rhetorical reference to Cihangir; in the words of Alpaslan and Bahadır, a more 'luxurious' neighbourhood in which 'theatre makers' and other intellectuals reside. Many of my interlocutors marked these differences in the educational levels and occupational profiles between the neighbourhoods.

Tellingly, this would at times intersect with narrations on 'marginal' lifestyles. In my interview with Ayşe, a 48-year-old Kurdish woman born in Bitlis, who moved to the neighbourhood with her family at the age of 6, she speaks of the 'marginal cafés' in the neighbourhood that, in her view, had replaced the neighbourhood practice of visiting the neighbour's home. Ayşe insists that every neighbourhood can be boiled down to 'clichéd' characters. For her, it is the 'tie-wearing cigar smoker' lacking *mahalle kültürü*, who spends his days sitting in the 'marginals' cafés' representing Cihangir. When asked about what happens at these cafés, Ayşe laughingly responds, 'They are working on their TV series' scenarios ... They destroy and reconstruct Turkey – but at the table! In their scenarios!' (interview, 25 November 2015). While thereby implying a politically oppositional stance on the part of these cultural producers, once more, this narration relies on lifestyle clichés tied to the adjacent neighbourhood of Cihangir.

Two female shopkeepers in the neighbourhood

Living in the neighbouring district of Şişli, which is governed by the largest oppositional party, the Republican People's Party (*Cumhuriyet Halk Partisi*, CHP), Aslı, a 32-year-old female entrepreneur, takes the metro each day to commute to Tophane. She works in the advertising industry, as does her female co-worker Beram, who is the same age. They had both aspired to produce a steady flow of income with the help of the café business. Aslı first moved to Istanbul from Izmir for her studies in 2013. Taking the place over from a friend in February 2014, Beram and Aslı soon learned they were paying three times the rent of what local shop owners were paying – only one of the reasons they felt like 'outsiders' to the neighbourhood. A little less than three years after its opening, they decided to close their café that had been located in the lower part of Kumbaracı Street, one of Tophane's new commercial arteries. Following a series of terrorist attacks in the city, as well as the 2016 coup attempt, their hopes that the economic situation of the café would soon stabilise, not least due to its proximity to Galataport, had vanished.

Both times I interviewed Aslı (interviews 22 July 2016, 21 January 2017), she described Tophane as being full of 'bigots'. The hairdresser's shop window just across from their café is adorned with a sticker showing the *Rabia* sign, in solidarity with the Muslim Brotherhood. The presence of Gönülbağı's office on the very same street made them feel the conservative presence even more strongly. Soon after they had first opened the café, several members of the local Muslim association visited it to impose several rules upon the two female café owners. They were told not to wear mini-skirts or any other 'revealing' clothing and not to sell alcohol. Being both surprised and angry at first, Aslı and Beram soon obeyed what they dubbed the 'rules of the *mahalleli*'s space'. According to Aslı, her female friends based in Cihangir would not come to Tophane to patronise her café, bothered by what they describe as the unambiguously punishing 'gazes' of the *mahalle*. While Aslı and Beram refuse to abide the angry *mahalleli*'s shouting at female tourists coming to their café to cover their legs, they did give in on another occasion. During Ramadan, their landlord asked them to close the café. While insisting on keeping it open to earn money for their rent, they agreed to cover the shop window with a cloth.

This compromise did not just serve their economic needs; the covered window also opened up a space of moral ambiguity, useful to both old and new users of the *mahalle*. One of the founding members of Gönülbağı, a 58-year-old former trainer for the local amateur football club, confesses that, due to his diabetes, he is not fit to fast during Ramadan. One day in June of 2016, he found shelter in Aslı's café. Sitting behind the covered window glass, he drank tea to cope with the heat of the summer, safely shielded from the looks of the *mahalle* (interview, 31 July 2017). According to Aslı, the *mahalleli* rarely make use of their space, not least for economic reasons: The tea at their café costs three Turkish Lira, which is six times the price of the tea served at the *kiraathane*, the traditional coffeehouse located just across the street. Subject to the economic pressure of a gentrified Tophane, spaces of consumption for these older *mahalleli* are increasingly vanishing.

A *de facto* café

An unusual place of consumption, it takes a second glance to recognise the teahouse on the ground floor of the dilapidated historical building located on Boğazkesen Street (Figure 11.2). I passed this informal *de facto* café many times, often seeing a big transporter vehicle parked in front of it. It belongs to some male family members from Siirt, who supply various shops in the neighbourhood. The inside serves as a place for socialising for them and many

Figure 11.2 Teahouse/AKP election information booth in Tophane.
Source: Photo by Urszula Ewa Woźniak, 2016.

other, oftentimes elderly, male residents of the neighbourhood. Furbished scarcely, the room looks like little has been spent on its décor for quite a while. The only object rendering this semi-public space a place of consumption is the Turkish-style double kettle placed on the counter of the kitchenette at the backside of the room. Yet, one wall almost fully covered with posters displaying Erdoğan and other leading party members of the AKP lends the room a slightly more formal quality. Having talked to some of the tea-drinking men sitting inside, I learn that the AKP had been using the space as an election information booth during election campaigns for a few years.

In October 2016, the users of this place were upset to learn that they would be evicted from the building. Notwithstanding the resistance of some, one autumn morning, the police violently dragged the last users out onto the street to lock the building. A well-known national holding company was said to have bought the estate to restore the building and open a hotel. A few months later, this *de facto* café's fate took an unpredictable turn: The former users of the café were able to return to it. With the holding's CEO now accused of being a putschist, the building's future remains unclear. The AKP's district office continues to use it for election propaganda; before any election, whether local or municipal, they send over posters that the users of the café then place on the same designated walls inside, thereby visibly marking the party's presence in the neighbourhood. Since early 2018, its former users, confirming that they are *mahalleli*, are glad to have been temporarily allowed back into the space. The rising rental prices in the neighbourhood made it impossible for them to rent a space nearby. Since they have been back, the rent has only risen a bit, Ali assures me in spring 2019. My question as to how long they will be allowed to stay remains unanswered. Despite their pronounced loyalty to the AKP and their claim to *mahalleli*-ness, the users of this space continue to be threatened with eviction. Having previously experienced that they are not immune to the lurking danger of displacement by the drastic socio-economic changes to the neighbourhood, their attempt to physically secure their place in the neighbourhood and thereby reinforce their imagined proximity to the hegemonic power of the nation also reflects their loyalty to the *mahalle* and its kin.

Conclusion

'Urban identity appeals to common memory and a common past but is rooted in a man-made place, not in the soil: in urban coexistence at once alienating and exhilarating, not in the exclusivity of blood' (Boym 2001, 76).

With its focus on old and new places of consumption, this chapter has discussed various formations of the *mahalleli* in the narrations of various users of the Istanbul neighbourhood of Tophane. Strikingly, in these narrations, the *mahalleli* mostly emerge as being male members of the constructed, territorialised neighbourhood community, ready to stand up for its moral order. As a construction of urban kinship, the notion of *mahalleli* thereby reflects a patriarchal order and implies a readiness to defend the *mahalle* – if necessary, with violent means. In the case of Tophane, these invocations of urban kinship clearly rely on the construction of an outsider community – that is, the adjacent neighbourhood of Cihangir.

The notion of the *mahalle* as an organic and pure community with clear-cut territorial boundaries reflects several anxieties. While the literature on encounters between different classes in the gentrified and thereby newly diversified *mahalle* spaces tends to focus on the anxieties of the urban middle class, the case of Tophane presented in this chapter sheds light on lower classes pleading allegiance to their *mahalleli* kin. Their anxieties revolve around the fear that the community of *mahalleli* will fail (to be): 'The neighbour who is also a stranger – who only passes as a neighbour – is hence the danger that may always threaten the community from within' (Ahmed 2000, 26).

Referring to a certain longevity within the *mahalle* and physical presence there, the *mahalleli* is constructed as the rightful and deserving resident of the *mahalle* – in moral, political, and socio-economic terms. Negotiations of belonging via the notion of *mahalleli* therefore reflect more than anxieties over the loss of lifestyles and moral values. Given the harsh distributional struggles over space and capital in view of ongoing processes of urban transformation, the case of the improvised café and AKP election booth is one among many micro-level contestations reflective of anxieties about livelihoods and economic and political positioning.

Negotiations of belonging to a specific urban *mahalle* are elementary to understanding both an academic and political discourse that has, when tied to discussions on urban modernity, long pitted urban spatialities against one another. The reproduction of the trope of a neighbourhood clash between Tophane and Cihangir reinforces long-known ideological contestations around Beyoğlu. While the encounters of various classes and their lifestyles in the space of the *mahalle* are certainly not new phenomena, in present-day post-coup Turkey, negotiations of being *mahalleli* are tied to questions of belonging to the nation. In this way, the *mahalleli* and the supposed 'neighbourhood culture' s/he embodies become highly politicised. Following the election of a CHP mayor in Istanbul after 17 years of AKP rule, it will be interesting to see whether this discourse of urban tectonics, which time and again almost inevitably causes frictions and eruptions, will continue to be reproduced.

Notes

1 Field notes, 20 July 2016. All translations are courtesy of the author.
2 During my fieldwork in Tophane, the members of a visibly dominant group of middle-aged men that sit in front of two hometown associations (*hemşehri dernek-leri*) sipping tea, weather conditions permitting, were often referred to as *mahalleli*. Strikingly, I later learned that many of these men do not in fact live in the neigh-bourhood anymore, but, due to the social fabric of the *mahalle*, come back to Tophane on a regular basis.
3 *Gecekondu* is Turkish for 'landed overnight'. The *gecekondu* neighbourhoods devel-oped as low-density informal housing areas for rural, migrant newcomers to the city.

4 By the early 2000s, Tophane and its surroundings had become the site of increasing construction activities. The opening of the Istanbul Modern Museum on the shores of the neighbourhood in 2001 triggered the movement of numerous art galleries to Tophane. More than a decade after the project's announcement in 2002, the harbour area became the centre of a grand-scale urban renewal project called Galataport, a public–private partnership between Doğuş Holding, Bilgili Holding, and the Beyoğlu Municipality.

5 In 2015, the district of Beyoğlu had a mean real estate value of 4,733 TL/sqm (Güvenç and Tülek et al. 2018, 63). When considering the individual administrative neighbourhoods that make up the *mahalle* space of Tophane, a more differentiated picture appears, showing that real estate values differ drastically across the neighbourhood, ranging between 3,791 TL/sqm and 10,782 TL/sqm (ibid.).

6 On the evening of 21 September 2010, the visitors of several art exhibition openings fell victim to a violent attack by a mob of around 40 men allegedly protesting against the visible consumption of alcoholic beverages on Boğazkesen Street, the main artery of Tophane. While this was neither the first nor the last incident of this kind in the neighbourhood, it was the one most noticeably debated in a wide range of media outlets.

7 Meaning 'the other side' in Greek, *Pera* refers to the other side of the Golden Horn. Beyoğlu was dubbed *Pera* in the Middle Ages.

8 All first names in this chapter have been anonymised.

9 Turkish for 'ruffian'.

10 German novelist Jörg Fauser depicted Tophane's infamously troubled past with drugs in the 1972 essay 'Tophane' and in his 1984 novel *Rohstoff*. Efforts to combat drugs in the neighbourhood continue today: In recent years, posters in the streets of Tophane have announced the fight against drug consumption. While the author of this campaign remains anonymous, many of my interlocutors assigned it to the local Muslim association Gönülbağı Vakfı.

References

Ahıska, Meltem. 2003. 'Occidentalism: The Historical Fantasy for the Modern.' *The South Atlantic Quarterly* 102 (2/3): 351–379.

Ahmed, Sara. 2000. *Strange Encounters: Embodied Others in Post-Coloniality*. New York: Routledge.

Aslan, Şükrü, and Tahire Erman. 2014. 'The Transformation of the Urban Periphery: Once Upon a Time There Were Gecekondus in Istanbul.' In *Whose City Is That? Culture, Design, Spectacle and Capital in Istanbul*, edited by Dilek Özhan Koçak and Orhan Kemal Koçak, 95–113. Newcastle: Cambridge Scholars Publishing.

Bartu, Ayfer. 1999. 'Who Owns the Old Quarters? Rewriting Histories in a Global Era.' In *Istanbul: Between the Local and the Global*, edited by Çağlar Keyder, 31–45. Lanham, MD: Rowman and Littlefield.

Bartu, Ayfer. 2001. 'Rethinking Heritage Politics in a Global Context: A View from Istanbul.' In *Hybrid Urbanism: On the Identity Discourse and the Built Environment*, edited by Nezar AlSayyad, 131–155. Westport, CN: Praeger Publishers.

Başaran, Pelin. 2015. 'Türkiye'de "Toplumsal Hassasiyetler" ve Tophane Vakası.' *T24*, May 28, 2015. http://t24.com.tr/haber/turkiyede-toplumsal-hassasiyetler-ve-tophane-vakasi,298019.

Behar, Cem. 2003. *A Neighborhood in Ottoman Istanbul: Fruit Vendors and Civil Servants in the Kasap İlyas Mahalle*. Albany, NY: State University of New York Press.

Bhambra, Gurminder K. 2007. *Rethinking Modernity: Postcolonialism and the Socio-logical Imagination*. Basingstoke: Palgrave Macmillan.

Bjarnesen, Jesper, and Mats Utas. 2018. 'Introduction Urban Kinship: The Micro-politics of Proximity and Relatedness in African Cities.' *Africa* 88: 1–11. doi:10.1017/S0001972017001115.

Blokland, Talja. 2017. *Community as Urban Practice*. Malden: Polity Press.

Bora, Tanıl. 2018. *Zamanın Kelimeleri. Yeni Türkiye'nin Siyasî Dili*. Istanbul: İletişim.

Boym, Svetlana. 2001. *The Future of Nostalgia*. New York: Basic.

Butler, Tim, and Garry Robson. 2001. 'Social Capital, Gentrification and Neighbourhood Change in London: A Comparison of Three South London Neighbourhoods.' *Urban Studies* 38 (12): 2145–2162.

Çavuşoğlu, Erbatur, and Julia Strutz. 2014. 'Producing Force and Consent: Urban Trans-formation and Corporatism in Turkey.' *City* 18 (2): 134–148. doi:10.1080/13604813.2014.896643.

Deeb, Mona, and Lara Harb. 2013a. *Leisurely Islam. Negotiating Geography and Morality in Shi'ite South Beirut*. Princeton, NJ: Princeton University Press.

Deeb, Mona, and Lara Harb. 2013b. 'Contesting Urban Modernity: Moral Leisure in South Beirut.' *European Journal of Cultural Studies* 16 (6): 725–744.

Dibazar, Pedram, Christoph Lindner, Miriam Meissner, and Judith Naeff. 2013. 'Question-ing Urban Modernity.' *European Journal of Cultural Studies* 16 (6): 643–658. doi:10.1177/1367549413497695.

'Erdoğan: Beyoğlu'ndaki marjinaller rahat durmazlarsa kulaklarından tutar fırlatırız.' *Cum-huriyet*, March 23, 2018. Accessed April 7, 2018. www.cumhuriyet.com.tr/haber/siya set/947611/Erdogan__Beyoglu_ndaki_marjinaller_rahat_durmazlarsa_kulaklarindan_tu tar_firlatiriz.html#.

Fortes, Meyer. 1969. *Kinship and the Social Order: The Legacy of Lewis Henry Morgan*. London: Routledge.

Gaonkar, Dilip Parameshwar. 1999. 'On Alternative Modernities.' *Public Culture* 11 (1): 1–18. doi:10.1215/08992363-11-1-1.

Güvenç, Murat, Murat Tülek, et al. 2018. 'Greater Istanbul and Istanbul Districts: Stratifi-cation of Neighborhoods with Respect to Age and Real Estate Values.' In *Services for Children and Family in Istanbul District Municipalities*, edited by Bürge Elvan Erginli, 63. Istanbul: TESEV Publications.

Hür, Ayşe. 2010. 'Tophane Kabadayısı, Beyoğlu Beyefendisi.' *Taraf*, September 26, 2010. Accessed June 12, 2016. www.arsiv.taraf.com.tr/yazilar/ayse-hur/tophane-kabadayisi-beyoglu-beyefendisi/13114/+&c-d=1&hl=de&ct=clnk&gl=tr.

Jackson, Emma, and Tim Butler. 2015. 'Revisiting "Social Tectonics": The Middle Classes and Social Mix in Gentrifying Neighbourhoods.' *Urban Studies* 52 (13): 2349–2365. doi:10.1177/0042098014547370.

Keyder, Çağlar. 1999. *Istanbul: Between the Local and the Global*. Lanham, MD: Rowman and Littlefield.

Lévy-Aksu, Noémi. 2013. *Ordre et désordres dans l'Istanbul ottomane (1879–1909)*. Paris: Karthala.

Massicard, Élise. 2018. 'Quand les civils maintiennent l'ordre. Configurations vigilantes à Tophane, Istanbul.' *Revue des Mondes Musulmans et de la Méditerranée*, April 4. http://journals.openedition.org/remmm/10212.

Mills, Amy. 2006. 'Boundaries of the Nation in the Space of the Urban: Landscape and Social Memory in Istanbul.' *Cultural Geographies* 13 (3): 367–394.

Mills, Amy. 2010. *Streets of Memory: Landscape, Tolerance, and National Identity in Istanbul*. Georgia: University of Georgia Press.

Mills, Amy. 2018. 'Cosmopolitanism as Situated Knowledge: Reading Istanbul with David Harvey.' In *Istanbul: Living with Difference in a Global City*, edited by Nora Fisher-Onar, Susan C. Pearce, and E. Fuat Keyman, 97–111. New Brunswick: Rutgers University Press.

Öncü, Ayşe. 2002. 'Global Consumerism, Sexuality as Public Spectacle and the Cultural Remapping of Istanbul in the 1990s.' In *Fragments of Culture: The Everyday of Modern Turkey*, edited by Deniz Kandiyoti and Ayşe Saktanber, 171–190. London, New York: I. B. Tauris.

Örs, İlay Romain. 2018. 'Cosmopolitanist Nostalgia: Geographies, Histories and Memories of the Rum Polities.' In *Istanbul: Living with Difference in a Global City*, edited by Nora Fisher-Onar, Susan C. Pearce, and E. Fuat Keyman, 81–96. New Brunswick: Rutgers University Press.

Öz, Özlem, and Mine Eder. 2018. '"Problem Spaces" and Struggles over Right to the City: Challenges of Living Differentially in a Gentrifying Istanbul Neighbourhood.' *International Journal of Urban and Regional Research* 42 (6): 1030–1047. doi:10.1111/1468-2427.12656.

Özyeğin, Gül. 2002. 'The Doorkeeper, the Maid and the Tenant: Troubling Encounters in the Turkish Urban Landscape.' In *Fragments of Culture: The Everyday of Modern Turkey*, edited by Deniz Kandiyoti and Ayşe Saktanber, 43–72. London, New York: I.B. Tauris.

Özyürek, Esra. 2006. *Nostalgia for the Modern: State Secularism and Everyday Politics in Turkey*. Durham, NC: Duke University Press.

Robinson, Jennifer. 2006. *Ordinary Cities: Between Modernity and Development*. London: Routledge.

Robinson, Jennifer. 2013. 'The Urban Now: Theorising Cities Beyond the New.' *European Journal of Cultural Studies* 16 (6): 659–677.

Rose, Nikolas. 1996. 'The Death of the Social? Re-figuring the Territory of Government.' *Economy and Society* 25 (3): 327–356. doi:10.1080/03085149600000018.

Soytemel, Ebru, and Besime Şen. 2014. 'Networked Gentrification: Place Making Strategies and Social Networks of Middle Class Gentrifiers in Istanbul.' In *Whose City Is That? Culture, Design, Spectacle and Capital in Istanbul*, edited by Dilek Özhan Koçak and Orhan Kemal Koçak, 67–92. Newcastle: Cambridge Scholars Publishing.

Tophanehaber. 2016. 'Cnn Türk'de Tophane ve Cihangir tartışması.' February 25, 2016. Accessed March 15, 2019. www.tophanehaber.com/cnn-turk-de-tophane-ve-cihangir-tartismasi/93/.

Türeli, Ipek. 2018. *Istanbul, Open City: Exhibiting Anxieties of Urban Modernity*. New York: Routledge.

Woźniak, Urszula. 2018a. 'Diversity in the Monochrome?' In *Grenzräume, Grenzgänge, Entgrenzungen: Junge Perspektiven der Türkeiforschung*, edited by Wiebke Hohberger, Roy Karadag, Katharina Müller, and Christopher Ramm, 63–82. Wiesbaden: Springer Verlag.

Woźniak, Urszula. 2018b. 'The *Mahalle* as Margin of the State. Shifting Sensitivities in Two Neighbourhood Spaces of Istanbul.' *Anthropology of the Middle East* 13 (2): 79–94.

12 The Basij of neighbourhood

Techniques of government and local sociality in Bandar Abbas

Ahmad Moradi

I'm not sure if I had a blackout. Constant coughing prevented me from seeing anything but a cluster of lights that seemed small, much smaller than normal. The heatstroke acted as a refracting lens perhaps. Lying on the ground, I saw the street filled with people, some of them bending over to examine my face. Restless feet and faces glaring with heat were bouncing up and down. I turned my eyes backwards, and the only things I could make out were tiny red and yellow headlamps flickering in the distance. We lost the football match to the alcohol dealers that night and I could see from the side of the street how discouraged Musa was. Musa, a bearded 20-year-old man and a cadre member of the Basij,[1] was our team's captain. After the defeat, his voice went on incomprehensibly, the voice of grave disillusionment. We lost the game and had to give up playing football from that night on, because nothing had gone as planned.

It was the third time we were playing football on a sweltering summer night in order to block illegal sales of alcohol on the main street of the neighbourhood. I could not bear the heat and almost fainted. The plan was to play football at around 8 p.m. every night, some minutes after evening prayer, firstly to find out who the clients were and, secondly, to stop them from getting to the end of the street to buy alcohol. In the midst of our game, the dealers forced themselves upon us by offering a deal. We were to play against them. If we won, we would be able to keep on playing on the street. If we lost, we would have to stop. We lost – which was not part of the plan. The plan was introduced by the Basij headquarters in the city to tackle the illegal transactions of alcohol in the neighbourhood with the help of the mosque-goers and the Basijis living in the neighbourhood. Football was supposed to be the third step of a four-phase plan that the commander of the base was told to carry out to get rid of alcohol distribution in the neighbourhood once and for all. I clearly remember Musa's cynical tone when he heard of the plan. The commander (*farmāndeh*) had just finished a meeting in the Basij headquarters and decided to impart the news to the members of the base. Musa was surprised, and I could see how his face fell at the news. The commander gave Musa a gentle laugh and resumed his talk by going through the breakdown of the plan:

First, we start by inviting the alcohol dealers to the mosque as an inquiry to find out what they need and why they have chosen this as their primary job. If this fails, we will talk to them one on one to persuade them to switch their job to the one we find for them. If that negotiation also fails, we have to intervene directly somehow. Probably we will play football where they sell alcohol. At least to show them we are watching them. And finally, if all the phases fail, we will report them.

The fatuousness of all this seemed very obvious as Musa replied,

We have done all these things. We even filed their names and sent it to the headquarters a couple of years ago. How does the headquarters expect us to find a normal job for them while they are making a lot of money every night?

It was the first time I had seen Musa slightly angry: 'When the police know exactly where these people are and help them get away with it, it means they are not determined enough or, God forbid, the police are with them.' Musa was sitting closer to the commander by now, trying to keep his voice as low as possible: 'What makes the dealers go on with their dirty job is the court that seems to have been bought by them. They afford their freedom by paying money to everyone in the judicial system.'

'I'm afraid to say that it's a command from the headquarters', the commander said. He didn't make any firm outcry, it was more of a cynically faint confession. Someone confided to me later that night that some years prior, the commander had set on fire the house of one of the dealers, who used to live in front of his house; court proceedings were still ongoing. This suddenly made all the debates surrounding countering alcohol transactions in the neighbourhood much more serious to me.

Despite their zeal to tackle social problems in their home community, many Basijis encounter frustrations as they attempt to implement dictated plans. Basijis like Musa are aware that these plans can be especially troublesome when it comes to the ways in which Basijis interact with locals. During my fieldwork in a poor neighbourhood in Bandar Abbas, a seaside town in the south of Iran, I encountered the different plans of the Basij, intended to guide and control how residents 'conducted' themselves.

Since the 1979 Revolution, almost all urban neighbourhoods in Iran have hosted a paramilitary base of the Basij. Known as the Basij of Neighbourhoods, they are the smallest operational unit of the Basij, where the presence and rhetoric of revolutionary ideology are put into practice. They were one of the first undertakings of the revolutionaries immediately after taking control of cities in 1979, and were envisioned to be central in policing the urban social order. In addition to acting as 'revolutionary bulwarks' (*sangarhāyih enghelāby*) against any security threats arising in each neighbourhood, in the years of the Iran–Iraq War (1980–1988), Basij bases played an active role in mobilising and recruiting volunteers for the front, and providing necessary manpower. In the post-war years, the Basij maintained its presence in neighbourhoods and its bases continued serving as places for extensive cultural, political, and military training.

The extent of the Basij interventions in the life of neighbourhoods varies and it distinctly targets men and women of different ages. Although the Basij recruits its members from both genders, men make up the majority of the membership. At the time of my fieldwork in 2015, there was no Basij base designed for women, as is common in other urban neighbourhoods, and training of the female members was limited to sewing classes in the mosque. These classes qualified participants for a certificate with which women could acquire low-interest loans and open tailoring shops. This situation made women less active in the organised activities that male members of the Basij were responsible for. A case in point is the holding of major national and religious celebrations and rituals in the neighbourhood. For example, the anniversary of the 1979 Revolution is a big ceremony to which the Basij invites speakers from other cities, decorates the streets with lights, and distributes confectionaries and juice from make-shift stalls. Muharram, and in particular Ashura,[2] is another important annual event during which religious mourning sessions are organised for ten consecutive nights.

Another significant responsibility of the Basij bases is to approve candidates' ideological qualification for a governmental position. The Basij plays a critical role in this procedure and informs institutions and organisations of candidates' commitment to Islam and the state. Such commitment is normally attested by the level of involvement in the Basij base and mosque activities. The same inspection process was conducted for university candidates, though was stopped in the late 1990s.

Given the wide range of their activities, the Basij bases in general serve as both structures of social control and places designed to recruit and train revolutionary forces. As a consequence of this situation, over the last four decades, the Basij bases have become places of intense social interaction, and places where different forms of engagement with the revolution are articulated and practised through governmental technologies[3] (Rose and Miller 1992).

Governmental technologies, as Rose and Miller define them, are 'the complex of mundane programmes, calculations, techniques, apparatuses, documents and procedures' which authorities utilise in order to 'shape the beliefs and conduct of others in desired directions by acting upon their will, their circumstances or their environment' (Rose and Miller 1992, 175). These technologies, they note, do not create an all-encompassing 'web of social control', rather they work through 'countless, often competing, local tactics of education, persuasion, inducement, management, incitement, motivation and encouragement' (1992, 175). I found the Basij involvement in organising local people's lives to be a significant arena for analysing a multiplicity of governmental interventions and for shedding new light on the flexibility and variations that characterise governmental practices in present-day Iran. I am interested in how members of the Basij subscribe to projects and ideas of revolution and attempt to translate them into practices within their neighbourhood, out of which they create a space for residents to engage in and experience the Basij's daily production of authority and governmental power.

Although, as I suggest, the presence of Basij bases represents the tangible exertion of governmental power over neighbourhoods, we would fall short of a complete picture should we neglect the ongoing compromises Basijis make by acknowledging neighbourhood relationships. For this reason, in this chapter I intend to show how techniques of government are interspersed with local techniques of sociality. In so doing, I follow Veena Das's approach, which recognises neighbourhood as 'the right scale', the 'analytical purchase [of which] comes from the force of its empirical character and not from it as an abstraction' (Das 2017, 191). As she explains,

> one place to locate the technology of government is in the practices followed by bureaucrats; such technologies, however, also have a life outside the offices of the bureaucrats as they are negotiated in other places such as in the low-income neighbourhoods.
>
> (Das and Walton 2015, 44)

Focusing on the neighbourhood, therefore, allows me to focus on the 'microscale' at which institutions of different scales intersect and are enfolded into neighbourhood relations (Das 2011; Han 2012), and where conflicts and negotiations of local groups are made and unmade.

Furthermore, by giving analytical weight to the 'neighbourhood', I explain how the revolution consists of ongoing efforts to manage distance and proximity in small social settings, and point to the ways in which different modes of engagement with the revolution are formed, maintained, and called for at the micro level. I argue that living together in neighbourhoods, with long periods of shared residence and shared histories, forces members of the Basij to attend to their existing relationships; the social relations of neighbours become a resource that is drawn upon and recycled in the making of an imagined collective group.

In what follows, I identify two examples of the Basij's governmental technologies at the neighbourhood level. First, I explore the concept of soft security and discuss how a revolutionary training programme aims to secure people's 'souls' as the prime site of crime prevention, control, and security intervention. In the following section, I focus on the use of numeric measurement in Basij bases, via which membership benefits are regulated. Finally, turning my analysis to the close entanglement of residents' lives with the Basij, in the final section I explore the concept of 'communities of complicity'. This concept, as Hans Steinmüller (2013) defines it, offers an analytical framework for examining the interplay of local sociality, official discourse, and state control.

Soft security: crafting the subject of prevention

Most residents of the neighbourhood I studied are Shi'a and speak a distinct Balochi language known as Koroshi. The neighbourhood is situated in the southeast of the city stretching over 198 hectares. Despite being reputed as one of the oldest inhabited areas in Bandar Abbas, the entire neighbourhood is catalogued as an informal settlement in official reports of the municipality. Informal settlements, in these reports, are defined as 'places hosting haphazard construction without compliance to urban planning and building regulations, and being crowded and polluted with minimum amenities' (Siahooi 2011, 30). Most informal construction in the neighbourhood occurred in the early years of the 1979 Revolution, during which migration from the countryside increased dramatically. According to the 2008 census, the population of the neighbourhood is 11,875 within 2,645 households. Unemployment stands at 14.2 per cent, one of the highest in the city. Taking into account recent economic stagnation and high inflation that has caused mass lay-offs, the rate of unemployment is likely to be much higher as of 2019.

The rampant social problems and the bleak prospects have made some residents, like Musa, determined to seek one seemingly viable solution: 'saving the youth' by training a new generation of forces committed to the revolution. Clarifying his commitment to a revolutionary training programme that he joined in 2012, Musa said:

> My ultimate goal in life is to turn this neighbourhood into a place which is mosque-centred and Basij-based. What made me a different person from my peers and saved me from the trap of addiction was the mosque and the Basij. So I want these children to do the same.

The programme, in effect, is part of a larger scheme for combatting the 2009 uprising. A month after the start of the popular upheaval, called the Green Movement by its supporters, the Basij and the Revolutionary Guards launched a nationwide revolutionary training programme called the 'Righteousness Circle'. It was designed for the 'persuasion' of Basiji members who had doubts about the value of the revolution and those who might have been affected by the questions posed by the Green Movement.

These circles are weekly sessions in which an experienced member of the Basij base, appointed by the commander, gives a 15-minute speech to a group of teenagers and encourages them to participate in discussions and to raise any questions they have on Islam or current political issues. The group I regularly participated in was divided into six subgroups and the head of each subgroup was responsible for keeping a record of attendance and ultimately informing the commander about what the members had been doing during the week. As an encouragement, with the permission of the head of each subgroup, a group of ten Basijis was taken to the swimming pool every week. Since the number of participants in the circles exceeded the number of Basijis who could go to the pool, subgroup heads needed to keep a close record of attendance and to be selective. This routine also acted as a system of punishment for those who had done anything wrong during the week. Once, I witnessed the head of a subgroup telling the commander to cross out the name of a member because he had stolen a headband from the base. It was also a rule that those who missed night-time prayers more than three times a week were not allowed to go to the pool.

Although Righteousness Circles aim to recruit and educate revolutionary forces, there is a local twist in this Bandar Abbas neighbourhood, where the Righteousness Circle also tends to provide a secure space that prevents teenagers from becoming alcohol dealers.[4] As Musa explained to me:

> This is one of the oldest neighbourhoods in this city and I was born and raised at the centre of it. Not that I'm overly sentimental about this neighbourhood, but it has always been reputed to have been at the forefront of political and social events. In 1979, people occupied the police station which is still some blocks away and joined the revolutionaries. It happened when no one in this city even knew that a revolution was underway. No need to mention the number of martyrs we had in the Iran–Iraq War. The neighbourhood is still at the forefront, but for completely the wrong reason: it has become the major distributer of alcohol in the city and the South. So it was very tough growing up here, considering that the nature of the neighbourhood is shifting from being a cradle of Muslim revolutionaries to a junkie-ridden, poverty-stricken and villainous place populated by youth who are either alcoholics or distributers of alcohol, or both. We really need to do something to prevent these children from becoming victims. Our attempt is to make the Righteousness Circles a response to this unfortunate situation.

Musa believes that, insofar as this programme makes being a Muslim and a revolutionary central to participants' lives, it prevents them from becoming alcohol dealers. In other words, the revolutionary training programme makes the work of alcohol prevention dependent upon the cultivation of an Islamic revolutionary self. This includes emphasising the role of the mosque in regulating people's lives, which creates a secure space and sufficient programmes for the Basijis to keep them busy at the base. Drawing on the work of Kevin Lewis O'Neill (2013) in La Paloma, Guatemala, which deals with the practices of American evangelicals to protect youth from joining gangs, I see the dual purpose of the revolutionary training programme as the creation of a 'subject of prevention'. O'Neill (2013, 205) defines the 'subject of prevention' as the 'individual imagined and acted upon by the imperative to prevent'. Using prevention as a pretext, the Basij, I argue, melds security interventions with revolutionary training, which is defined by the Basij as 'soft security' (*amnyat-e narm*). It is the idea of soft security and the practice of prevention that is the main focus of this section.

After the commander's discussion with Musa about the plan for fighting alcohol, it took them several days to convince people in the mosque to gather for the meeting. It was supposed to be the first phase of the plan. None of the dealers showed up at the meeting and it became an occasion for the mosque-goers to exchange their opinions on this issue. They were in complete disapproval of the plan, as it was quite clear that the dealers would not show up at the mosque to seek help and freely divulge their circumstances and needs, not to mention publicly admit to their 'crimes'. Some even objected that dealers are alcohol consumers and should therefore not be allowed in the mosque; they accused those who designed the plan of failing to even consider obvious Islamic norms. In the midst of scattered objections and a cacophony of whispers, the commander asked a man who was participating in the meeting on behalf of the Basij headquarters to give a talk on the future plans for the mosque and the Basij base and to further explain what the commander meant by the 'operational plan' (*barnāmeh amaliyaty*).

'Rather than choosing a threat-oriented analysis (*tahdid mehvar*), we need to adopt an opportunity-oriented analysis (*rasad mehvar*).' He started his speech by proposing a necessary shift in the security practices of the Basij – a shift that marks a new tendency in the Islamic Republic to choose soft measures for counteracting social problems. He saw the plan for the eradication of alcohol transactions in this light. However, in his definition of opportunity-oriented analysis, he seemed to be more concerned with preventing the youth from becoming involved in alcohol transactions than with dealing with those who were already involved: 'To train our forces, we first need to create a space that is attractive for the youth. The same space in mosques and the Basij bases that we had in the early years of the revolution.' He continued:

> We have realised that we need to conglomerate our efforts and give a direction to revolutionary forces. We call this the operational plan. Within this framework, we first build our own forces, and then we do the cultural feeding (*khorāk-e farhangi*).

His remarks demonstrate how the office seeks a unifying power to coordinate the revolutionary practices. These practices are deeply bound to the construction of new subjects whose very minds are to be secured against dispersion and must be attuned to a single centre. It is this orientation to the centre that is claimed to have been absent from the life of revolutionary forces prior to the 2009 uprising:

> It must be emphasised that we cannot advance the revolution with pious people. We don't need people who believe in all the principles of piety but are not revolutionary. We need to orient all the practices of the mosque towards revolution and then be pious Muslims and believers. This will pre-pare us so that we will have revolutionaries at critical moments. But before that, we need to have a comprehensive package with which to strengthen our forces, strengthen their thoughts and respond to their questions.

During his speech, he discriminated between two forms of practice: a practice that is part of the larger scheme of realising revolutionary values and Islamic culture in the entirety of one's life, and a practice that is Islamic in form and style but does not necessarily serve to train and realise a revolutionary self. Drawing on such discrimination means that the form of revolutionary self that the Basij office seeks to realise is predicated upon the transformation of many aspects of social life. This not only favours restoration of society over individual reform, but also prioritises revolution over piety. The desire to restore Iranian society to a point in history experienced in the early years of the 1979 Revolution is not entirely new, as it has been constantly pursued and expressed in the past decades by various groups. However, this presumed duty has become markedly prominent for the Basij and the Revolutionary Guards in recent years as they have been entrusted by the authorities with increased responsibilities in the aftermath of the events of 2009.

> To achieve our goal in establishing a genuinely revolutionary country, we first need trained forces who are keen to act revolutionarily. We have problems in all cultural spheres: family, TV, neighbourhood, alcohol consumption and so on. In the meantime, in this mosque and its Basij base, we have had a successful track record in recruiting forces. We encourage you to take the young members of the Basij to the park, pool and even go shopping with them. We want these children to feel your presence everywhere in their life. See every action of yours for these children as an operation and try to evaluate all those operations among yourselves. This is what we call soft security.

His talk shows the coupling of the revolutionary training programme to intimate forms of security intervention. In this marriage, a new method, soft security, has emerged that attaches importance to intimacy in building new revolutionary forces. This allows me to make two interlaced arguments.

First, soft security, which does not rely on hard measures and which prioritises long-term training over temporary mobilisation, works through a distinction that divides residents of the neighbourhood into two camps: Islamic revolutionaries and the rest. For this reason, revolutionary training programmes are selective and involve imperfect accommodations. In attempting to craft the subject of prevention and to carve out a social space where members of the Basij are closely monitored and kept safe, programme priorities in selected areas entail the neglect of those in other areas. As Kevin O'Neill (2013, 219) writes in reference to evangelical mentoring programmes in Guatemala, the imperative to 'prevent' gives rise to 'fields of perceptibility [that] make individuals legible (or not); they make certain people needed (or not)'. I will explain this further in the next section by detailing the hierarchies of membership in the Basij.

Second, the Basij's prescribed programmes for training revolutionary forces through practices of prevention are mainly built on the idea of long-term engagement with the lives of teenagers and close observation of their routines. In a sense, these efforts at generating intimacy render intimate relations between neighbours instrumental in creating the ideal model of revolutionary subjects. Neighbourhood, from the perspective of the Basij headquarters, serves as an idealised conception of a collective group and a set of values, resting on and perpetuating ideological and intimate relations. Nevertheless, as the next section shows, this type of relationship, qualified by neighbourhood intimacy, is characterised more by agonistic intimacy (Singh 2011), i.e. marked by tensions and unequal relations.

Extending my analysis in relation to the two points I have highlighted above, in what follows I explain how neighbours are enmeshed in close relations with the Basij, whether through spatial intimacy or networks of service relations, and how their experience of and encounters with the Basij base are shaped around contested linkages between forms of intimacy on the one hand, and governmental techniques of bureaucracy on the other.

Measuring faith: creating the space of negotiation

As evidenced in the previous section, the Basij strives to set normative expectations for public participation at the level of neighbourhood. In practice, these expectations shape and reinforce the dominant moral order, which is backed by state institutions and aided by bureaucratic powers of administration. Nevertheless, I argue that despite adherence to the dominant moral regime, Basijis are implicated in a different moral order than residents of the neighbourhood, which makes them accountable to their neighbours and demands appropriate responses to their neighbours' ethical claims.

This situation reveals how members of the Basij simultaneously subscribe to competing moral orders in which corresponding rationalities and values are largely incompatible. In this section I focus on ongoing negotiations between individuals and local state officials and show how competing moral orders produce differing normative expectations regarding practices of being a good neighbour and a responsible state official.

In every Basij base in the country there is an office that deals with membership. Once registered, you become a regular Basiji. However, if you intend to benefit from the membership in the form of loans, a reduction in months of military service, jobs, and so on, you have to go through special training and become an active member, which requires almost six months, providing that you take the courses intensively. Otherwise, the process can take years and often ends in failure. Therefore, one of the main tasks of the Basij bases is to keep a record of members' activities. In 2013, the Basij introduced an online system for recording the courses that members take. This lets the main Basij office closely monitor the practices of its members. Within this section of the base, the 'human resources office', I could see how neighbourhood residents seeking certification of their membership would negotiate their status. They were mainly regular members who did not frequently participate in the base's programmes and activities. The office gave me an opportunity to talk to these people seeking membership certificates for the benefits they bring.

The commander was busy filling out some forms that needed to be sent out to the Basij headquarters. The commander asked me to sit in the human resources office so he could show me some videos taken on their recent pilgrimage to Mashhad, the holy city in the northeast of Iran. The form he was filling out was an assessment sheet on the basis of which regular Basijis are upgraded to active members. At the top, it read 'Ideological Inquiry Sheet for Upgrading Membership'. I asked the commander to let me have a glimpse. With some hesitation, he gave me a blank sheet, telling me that the completed sheets were confidential. What surprised me the most were the columns with some numbers underneath, indicating that the commander had to rate the Basijis using the numbers. However, the headings dealt with rather abstract and immeasurable characteristics: On a scale of 8–20, evaluate the belief of the respected member in the basics and principles of Islam; belief and practical commitment to the Supreme Leader (8–20); commitment to prayer (6–15); participation in religious and political ceremonies (4–15); observance of Islamic morality and Basij codes of conduct (4–15); selflessness and readiness for sacrifice (6–15).

The last column, on selflessness, made me start questioning the concreteness of such forms, as it seemed far removed from what I had expected from the assessment sheet. The commander disagreed with the whole procedure, showing no faith in the forms. 'Then why do you think they designed the form like that?' I asked him. He replied:

> The Basij has turned into a huge bureaucratic organisation and needs to have a system of evaluation. Nevertheless, the organisational aim is to direct the hearts and minds of the members because it believes that, when you are persuaded enough about an issue, it is engraved in your soul and you act accordingly. We didn't used to need such a system – everyone was a believer in revolutionary values and Islamic principles; now things have changed.

Our dialogue was interrupted by the arrival of two young men in the office. One of them was the same man I had seen during our football match with the alcohol dealers some nights ago. He was later introduced to me as Black Davood. He had an unforgettable face, his forehead, in particular: On that dark face was a forehead covered with crusty patches of silvery scales; it was like staring at a sandy beach glittering at night when the sea had retreated. As I later found out, he used to be a professional wrestler but, because of his skin disease, had given up sport.

He was with a short man in his early twenties. His friends called him Tidy Saeed. He walked with a limp and, as soon as he sat down on the seat arranged in front of the commander, the commander asked him to tell us when he had been released from prison. Shyly, he explained that it was a mistake and that he had not stolen the motorcycle – it was just an allegation that had never been proven. 'That's why you were in prison, ha? Just for doing nothing', the commander said scornfully, raising his voice. With his head down, Saeed mumbled some words. Davood intervened, perhaps to lighten the suffocating atmosphere of the room: 'He has changed. He wants to do military service and find a job afterwards. Would you just give him a certification for active Basijis so that he can reduce the months of his compulsory military service?'[5]

I was still holding the assessment sheet in my hand and was contemplating what score Saeed would get. 'No way can he pass the evaluation', I thought to myself. 'Let me check your profile on the computer', the commander said, as he turned his chair and logged in. After a minute, he turned his head towards Davood and Saeed with a big grin on his face:

> What do you expect me to do for you kids? Saeed, you never finished a course on *vilāyat-e faqih* ('the Governance of the Jurist') nor did you participate in a session of Righteousness Circles. Plus, you have a criminal record.

SAEED: Nothing you can possibly do? Like, just sign the paper. Who would know?

COMMANDER: It's all registered on the Basij central computer. Nothing I can do, I'm afraid.

SAEED: So how long should I come to become an active Basiji?

COMMANDER: If you pass the courses and then military training. So almost five months.

SAEED: I want to have four months' reduction from my military service and for that I have to come here every night for five months. What person with any common sense would do that??

COMMANDER: It's up to you Saeed. And don't fool yourself, please. You have a criminal record. You could never pass the ideological interview at the end.

Disappointed, but with some humour, they left the room. It seemed as if they had already been sure that they would not get the paper and had just come to try their luck. Some days after, I saw Saeed on the street and it gave me an opportunity to engage him in conversation about what had made him decide to join the Basij.

Saeed started,

> I was at middle school when a classmate asked me to go on a school-trip to Mashhad with him. We didn't have to pay anything, so I decided to go. After the trip, I visited the Basij base. The base was close to our house anyway and I could visit my friends there at the base. I went to the base for more than a year, till my elder brother got arrested. When I visited him in prison, he was very angry at me. I realised that what I had told the commander in confidence had been used against my brother during his interrogation.

'How did your brother know that it was you that had revealed his secrets?' I asked.

'He knew because he was always telling me not to hang out with the people at the base. Plus, what they knew was quite personal. It was only me who knew those things, and the commander.'

'So you stopped going to the base after that?'

'No, I went to the base a couple more times. But I stopped going when I had a huge fight with the commander two years ago. You know I couldn't really tolerate his betrayal.'

Some days later, I spoke with two adult members of the base. One was responsible for keeping records of the Basijis and minutes of the Tuesday meetings for base members. He defined his role at the base as the 'inspector': someone who keeps records of meetings and files relating to the Basijis. The other man was someone I generally met at the base while he was taking teenagers to the pool. Our conversation was about finding ways to increase the number of members at the base. He was complaining that he had had enough of doing voluntary tasks, coming to the base every night and not being able to take care of his family. 'If we had more members, we would be able to share the duties and spare more time for ourselves', the inspector said.

I thought it might be the moment to share with them the interaction I had seen between the commander and Tidy Saeed the other night, to find out what they thought about the Basij's strictness in upgrading members despite its claim of being a voluntary organisation whose door is open for all Iranians to participate.

The inspector replied,

> I was Saeed's classmate in high school. He is not a type of person that does voluntary duties. But you know Saeed wants to have the benefits too. Like two weeks ago, Saeed called me to say that he needed my help to get him out of a tough situation. His sister, who got divorced over a year ago, was threatened by her ex-husband, who said that she couldn't marry anyone else, just days before her second marriage. The ex-husband entered their house and started beating up Saeed's sister. He was so angry that he might have killed her. When Saeed called me, he was desperate and he was breathing heavily. I ran to the base and collected some people and the commander. We were at Saeed's place less than 10 minutes after he called. The commander showed her his Revolutionary Guards card and warned the ex-husband that if he came anywhere near the neighbourhood again, he would get into trouble and would have to deal with Revolutionary Guards intelligence. You can't believe how sheepishly he left Saeed's place. How do you think Saeed could solve his problem so effectively?

As this vignette shows, how virtues are quantified and the performance of Basijis is audited creates a hierarchical structure that necessarily leaves some people behind. However, this hierarchy does not mean that ordinary people are completely left out. Through interactions with the Basij, regular Basijis create their own space of negotiation, in which they strive to gain more of the benefits facilitated by the Basij. As a result, hierarchy of membership serves two purposes: It enables local officials to enter the personal lives of local people and it calls on individuals to embody and internalise the revolutionary ideals that the regime seeks to inculcate in them. In so doing, it also defines a broader conception of engagement – one that is shaped by loyalty and forms a basis for negotiation. It is the latter that forms the crux of my discussion here, especially in relation to the ways that the daily interactions between citizens, like Davood and Saeed, and the Basij base produce spaces of negotiation. The combined effect of auditing practices and the imperatives of neighbourhood intimacy raises questions about the nature of the Basij's bureaucratic practices and neighbourhood intimacy.

Attention to bureaucratic practices and various articulations of and engagements with these practices helps us think through the complex relationship between the Basij as a national organisation and a local actor. Staff of the Basij base simultaneously articulated a sense of being responsible for their bureaucratic role and expressed their dissatisfaction at how the Basij had become institutionalised. This means that the staff are uncomfortably positioned between bureaucracy as an ideology and practice of accountability (Herzfeld 1992), on the one hand, and local sociality on the other. This feeling was pervasive in all the Basiji groups that I worked with; nonetheless, it was most evident at Basij bases. Unlike other headquarters, the Basij of Neighbourhood base operates at the local level, with various kinds of intimacies. As mentioned towards the end of this section's vignette, the inspector and his colleague knew Saeed and Saeed sought their help because of their neighbourly intimacy.

I should emphasise that I do not see intimacy as a form of bonding, without conflict. I argue that paying attention to neighbourhood relations can be analytically productive when we consider that the intimacy existing between neighbours is upheld by 'practised relatedness', and that it is also troubled by, but endures, through conflicts. To grasp the complexity of neighbourly relations, Singh (2011, 430) develops the concept of 'agonistic intimacy' as a way of capturing the co-presence of conflict and cohabitation. As defined by Singh, the term offers 'a picture of relatedness, whose coordinates are not predisposed entirely towards either oppositional negation or communitarian affirmation' (Singh 2011, 431). The coordinates of the figure of neighbours in her ethnographic description are not those of 'potentially hostile neighbouring groups' (Singh 2011, 430), but groups who are involved in demotic and local tensions that every neighbour could be vulnerable to. This is most evident in Saeed's reliance on the Basij base to be saved from desperate circumstances, despite his awareness that the same people in the base had informed on his brother and betrayed his honesty.

Yet, as I show in the next section, there are instances in which neighbours direct their antagonism beyond the community, while the shared intimacy of neighbours helps them resolve tensions through irony and complicity.

Communities of complicity: creating the space of contention

The social and political landscapes of neighbourhoods reproduce specific power relations and reveal the many ways in which the increasingly complex interests of individuals, families, and groups interact and clash with the goals of the state. In this section, I argue that the multiplicity of social and political relations not only fosters conflicts, but also shapes forms of loyalty and strategic alliances. As a case study, I focus on the unfolding of a labour strike in the wake of mass lay-offs in an aluminium factory in Bandar Abbas, during which the lives of some residents of the neighbourhood where I conducted fieldwork were affected. Tracing the way in which the labour strike evolved over a week, I show how the Basij of Neighbourhood episodically responded to the strike: from direct confrontation to becoming complicit with labourers in forging documents. Through these episodes, I highlight how neighbourhood loyalties, which are deeply embedded in intimate relations, outweigh other loyalties – to the state or the revolution – and at times may stand in stark contrast to them. Moreover, I show how these loyalties and tensions were strategically called upon by residents to make their claims and frame their grievances: first, by framing their arguments with the dominant political discourses and, second, by relying on neighbourhood relations and their intimate lives.

To explain this situation, I draw on Hans Steinmüller's (2013) work, which deals with the everyday life of 'official' discourses in vernacular processes, processes within which Chinese villagers and local officials use irony to express the differences between the 'official' version of what it is to be a Chinese peasant or a 'good official' on the one hand, and the 'pragmatic' actualities, i.e. life as it is actually lived, on the other. Steinmüller explains that this process is deeply informed by a shared sense of intimacy between villagers and local officials. Through this intimacy, these two groups constitute what he calls 'communities of complicity'.

I first heard about the strike at Musa's house. Three men who lived in the neighbourhood had stopped by to give him news about the recent labour strike. From their conversation, I understood that the three men had been laid off a week before and that they believed that the same would happen to many more labourers in the days to come. While on strike, they encountered some staff from the Basij of Neighbourhood, along with riot police, who were deployed to confront the workers' action. 'They were not violent, but they threatened over and over that we would be in trouble should we continue with our strike', one of the men, who seemed to be in his late fifties, said.

The aluminium factory was established in 1990 and was owned and run by the state. It was handed over to the Basij in 2015, according to a privatisation scheme that involved the transfer of state-owned companies to 'private entities'. Faced with debt and low production, the Basij, as the new owner of the factory, claimed that they urgently needed to cut down the number of workers, or else they would have to file for bankruptcy. This announcement sparked the labour strike in the factory. On the same day, workers took part in the walkout and later blocked the main road into the city.

As continuing the road blockage was too risky and strikers were likely to end up in prison, workers came up with an alternative plan. Carrying photos of the Supreme Leader, more than a hundred striking aluminium workers gathered in front of the Basij headquarters in the city. They called on commanders of the

Basij to intervene by pressuring their colleagues running the factory to pay the delayed salaries. They also demanded the resignation of the factory's managers and the return of factory management to the state. They marched down the street chanting: 'Down with the labourer, long live the oppressor' (*marg bar karegar, doroud bar setamgar*); 'The poor labourer should be executed, corrupt businessman should be released' (*karegar-e bichareh e'dam bayad gardad, mofsedeh eghtesadi āzād bayad gardad*); 'Salaries are not paid, down with the USA' (*hoqoqha rā nemidand, marg bar America*).

Laid-off labourers made strategic use of official revolutionary discourse in order to secure their jobs, and more broadly to make legitimate claims that the revolutionary organisation of the Basij could not overlook. By ironically turning famous revolutionary slogans on their head, workers appropriated language that is familiar to both the Iranian public and above all to the Basij.[6] By retooling the official rhetoric, the laid-off labourers – turning against their employers – expressed their dissatisfaction and pressured commanders for not standing by the revolutionary promises of inclusion of the poor. In this context, workers used hegemonic language to frame their contention, to justify their grievances, and to demand that the revolutionary organisation of the Basij translate its revolutionary rhetoric into concrete practice.

Five days after the strike began, the factory's managers, appointed by the Basij, promised to pay the delayed salaries in full and to provide senior laid-off workers with retirement schemes, which could be claimed from the Social Security Organisation (SSO). The insurance coverage was supposed to be arranged in a couple of months. Striking workers agreed and decided to temporarily halt their strike. The next day, however, after consulting the SSO, senior workers realised that, for the promise of the retirement scheme and insurance to come through, they needed an official letter from the managers stating their agreement with the arrangement within two days. After the implementation of a new labour law, employers were no longer allowed to authorise provision of such insurance. The managers refused to sign the letter in less than two months. The dispute escalated again, but this time with less force, because junior workers did not join in, as they considered senior workers to be fighting only for their own rights and ignoring their plight.

The next day, I encountered a scene at the Basij base which is crucial to my argument in this section. After a few minutes, I realised that some senior workers had crept through a now defunct factory's warehouse after midnight the night before, found their way into the managers' office and stolen the official stamp and some sheets of letterheaded paper. Now, they were asking Musa to write the letter that the managers had refused to issue; they would later sign and stamp it, before handing it in to the SSO.

They repeatedly said that they were not sufficiently literate to write the letter, so they asked Musa to write it on their behalf. Musa remained unmoved in his chair, hunching over the papers while they continued talking. The workers, however, did not seem to be people who would give up easily; they just hovered there, and started to spew insults at the corrupt politicians and commanders of the Basij. Angry now, Musa pushed the chair back and crossed the room to one of the empty chairs against the far wall behind the drawers. The workers followed him. One of them kicked the side of his chair so hard it nearly sent the papers flying. Suddenly noticing how tense their interaction had become, the workers exhorted the man who had kicked the chair to control his temper and reassured him that Musa was on their side.

Soon after, Musa glanced up, looking distractedly at the workers lined up in the room. He then put the paper in front of him and started writing the letter. It took Musa quite a long time to finish the draft; crumpling up the paper twice, he wrote and franticly crossed out the sentences he had left incomplete. While he was writing, the workers encircled Musa's chair and kept saying to him, 'This is the right move, Musa. You're not one of them'; 'You're proving who a true revolutionary is'; 'This is what the Basijis stand for (*in Basiji hast ke ma mishināsim*), not those commanders getting fat by devouring the poor's share'. When Musa finished the draft, he read it aloud and passed it to one of the workers, 'I'm not going to sign or stamp it. You should take the whole responsibility for the letter. Is that clear?'

When the workers left, with the paper in their hands, only Musa and I stayed behind at the Basij base. But I kept wondering how Musa felt about his complicity in forging the documents. We remained silent for a while and I pretended that I was busy arranging the chairs. Suddenly, Musa said

> How could I look them in the eye the next day if I refused their request? How am I going to talk about revolution to the children of these fathers, if I turn a blind eye on the injustices (*zulm*) that are waged against these poor people?

I find the concept of 'communities of complicity' useful to account for how workers and local members of the Basij actively negotiated the tensions between Basij-propagated discourses and local sociality – the kind of tensions that bind these two groups together. In Steinmüller's (2013) account, local officials are not mere representatives of the state, but an essential part of the 'communities of complicity'. He notes that local officials and villagers share an intimate knowledge of and familiarity with local practices, and that 'the complicity they share is based on shared experience in local sociality on the one hand and on an intimate knowledge of the condemnation of some elements of it on the other' (Steinmüller 2013, 94). In his ethnography, Steinmüller shows that local officials experience embarrassment as they take part in officially condemned practices such as 'favours' or 'gambling'. Viewing embarrassment as an indispensable part of communities of complicity, Steinmüller makes use of Herzfeld's concept of cultural intimacy as 'the recognition of those aspects of a cultural identity that are considered a source of external embarrassment but that nevertheless provide insiders with their assurance of common sociality' (Herzfeld 2005, 3). In this sense, what local practices have in common is 'the fact that their outside representation is overwhelmingly negative, whilst they are really essential in everyday life' (Steinmüller 2013, 219).

In the case of striking labourers and Musa, however, embarrassment was not an issue. This is where my take on complicity differs from that of Steinmüller. If Steinmüller defines cultural intimacy as a constitutive part of communities of complicity, the complicity of the Basij and the workers is rooted in the shared revolutionary aspiration of justice. In other words, if complicity in the case of Chinese villagers is revealed to and concealed from the outside world through gestures of embarrassment, irony, and cynicism, in my case it is an explicit gesture of contention against the outside world and beyond the community. This gesture of contention is practically rooted in the intimate relation between the workers and the Basij of Neighbourhood and reveals the limit of the moral order imposed from outside the community. In fact, the formation of this 'community of complicity' provides an insight into, first, how the interplay of local practices and official discourses produces or fosters contention (as discussed in the previous section through the concept of agonistic intimacy). Second, it reveals how members of the Basij actively and reflexively deal with the complicit position of the Basij within the system that is the object of criticism. And, finally, it shows how the shared intimacy of neighbours outweighs the formal and imposed official discourse and practices. Faced with competing moralities, neighbours favour the daily interaction over what seems to be impersonal modes of rule.

Conclusion

Neighbourhoods present a rich terrain for ethnographic inquiry into the conflictual nature of 'collective life' in cities. Taking up neighbourhoods as a 'unit of analysis', ethnographic research is especially well suited to examine ordinary urban lives, revealing the abiding tensions in 'micro-geographies' and the way in which collective life is imbued with conflict. Rechtman (2017, 134) argues that 'the real collective level of lives as lived [in neighbourhoods] is not a construction; it is a fact'; neighbourhoods provide a shared space where residents breathe the same air, experience similar events, and struggle through similar contradictions. Indeed, it does not mean that neighbourhoods are a bounded entity, or a homogenous unit. There are multiple actors, with different ideological persuasions and varying degrees of connection to the institutions of power, who participate in governing neighbourhoods, and collectively shape the life lived.

In this chapter, I focused on the Basij as one of the key actors in Iranian urban settings. The sheer number of Basij bases in neighbourhoods and the extension of their interventions have given the Basij a prominent role in regulating collective life at the neighbourhood level. Such a role is substantially reinforced by the distinctive technologies of control they employ, examples of which being training programmes and hierarchies of membership. Nonetheless, we fall short of giving a complete picture should we stop at describing the Basij of Neighbourhoods only as mechanisms of social control.

I have argued that, while the Basij continues to exert its power over the residents, the members of the Basij increasingly appear to locals as low-level state functionaries who are there to serve as the visible face of the state and to redress local grievances. I discussed how the Basij of Neighbourhood is involved in 'administrative and hierarchical rationalities that provide seemingly ordered links with the political and regulatory apparatus of a central bureaucratic state' (Das and Poole 2004, 5). Paying close attention to these 'parochial sightings of the state' (Das and Poole 2004, 6) allows us to see how Basijis, such as the commander, are ambiguously located between local notions of sociality and an impartial bureaucracy. In other words, how they appear are as low-level officials who acknowledge formal rules but are obliged to recognise the importance of personal relationships.

This presents us with a chance to go beyond viewing the Basij of Neighbourhood only as an ideological space for the reproduction of revolutionary rhetoric and practices in the interests of the regime (Golkar 2012, 2015; Bayat 2013). Rather, I examined substantive lines of disputes, conflicts embedded in everyday life, and the intimate relations of neighbours. Although I do not wish to downplay the dominating power of the Basij, I emphasise the need to examine on-the-ground understandings of 'techniques of government' and 'social relations' – and the flexibility of the boundaries between them.

Notes

1 *Sāzmane Basij Mostazafān* (The Organisation for the Mobilisation of the Oppressed), or, for short, the *Basij*, is a large and highly significant pro-regime paramilitary organisation in Iran. It came into existence at the command of Khomeini in 1979, shortly after the end of the Pahlavi regime. Four decades later, the Basij claims to have over 15 million male and female members, over 18 per cent of Iran's population of 80 million. It is estimated that there are over 35,000 Basij bases spread across every city and village in Iran (Jalili 2014).
2 Ashura is the tenth day of Muharram, when Imam Hoseyn, the third Imam of Shi'a, was martyred in the Battle of Karbala in 680 CE. Shi'a think of Ashura as the key historical moment in their history and interpret it in terms of the necessity of revolting against oppressive rulers and of fighting for one's faith and beliefs.
3 Rose and Miller draw on Foucault's notion of government as 'the conduct of conduct', or 'the way in which one conducts the conduct of men' (Foucault 2008, 186). Government, in Foucault's view, is a pervasive activity carried out by a wide range of bodies, including but not limited to state elites, which can be traced both in small fractions of the society or in the social body in its entirety, and involves practices, techniques, and rationalities aimed at managing the population. In Foucault's definition, government may be 'understood in the broad sense of techniques and procedures for directing human behaviour. Government of children, government of souls and consciences, government of a household, of a state, or of oneself' (Foucault 1991, 81). Drawing on the same analytical framework, this chapter explores governmental practices of the Basij at a neighbourhood level.
4 The distribution and consumption of alcohol is strictly prohibited in Iran, but alcohol is smuggled to the country through the borders. Located by the sea, Bandar Abbas

has been one of the major ports into which alcohol is smuggled. Having access to boats and travelling to the Gulf countries, fishermen in this neighbourhood are particularly prone to being involved in smuggling alcohol. In recent years, as unemployment has reached unprecedented rates, more locals have taken up smuggling alcohol as their source of income.

5 Military service is compulsory for men in Iran and lasts for almost two years.

6 Irony and humour as 'acceptable forms of confronting authority' have been extensively explored in anthropological literature (Fernandez and Huber 2001; Boyer and Yurchak 2010; Klumbyte 2011). For a specific discussion on ironic slogans, see Haugerud (2010), Knight (2015).

References

Bayat, Asef. 2013. *Life as Politics: How Ordinary People Change the Middle East.* Stanford, CA: Stanford University Press.

Boyer, Dominic and Alexei Yurchak. 2010. 'American Stiob: Or, what late-socialist aesthetics of parody reveal about contemporary political culture in the west.' *Cultural Anthropology* 25 (2): 179–221.

Das, Veena. 2011. 'State, citizenship, and the urban poor.' *Citizenship Studies* 15 (3–4): 319–333.

Das, Veena. 2017. 'Companionable thinking.' *Medicine Anthropology Theory* 4 (3): 191.

Das, Veena and Deborah Poole. 2004. 'State and its margins: Comparative ethnographies.' In *Anthropology in the Margins of the State*, edited by Veena Das and Deborah Poole, School of American Research advanced seminar series, 3–33. Oxford: James Currey.

Das, Veena and Michael Walton. 2015. 'Political leadership and the urban poor: Local histories.' *Current Anthropology* 56 (S11): S44–54.

Fernandez, James and Mary Taylor Huber, eds. 2001. *Irony in Action: Anthropology, Practice, and the Moral Imagination.* Chicago: University of Chicago Press.

Foucault, Michel. 1991. 'Governmentality.' In *The Foucault Effect: Studies in Governmentality*, edited by Graham Burchell, Colin Gordon, and Peter Miller, 87–104. Chicago: University of Chicago Press.

Foucault, Michel. 2008. *The Birth of Biopolitics: Lectures at the Collège de France, 1978-1979.* New York: Palgrave Macmillan.

Golkar, Saeid. 2012. 'Organization of the oppressed or organization for oppressing: Analysing the role of the Basij militia of Iran.' *Politics, Religion & Ideology* 13 (4): 455–471.

Golkar, Saeid. 2015. *Captive Society: The Basij Militia and Social Control in Iran.* New York:Woodrow Wilson Center Press, Columbia University Press.

Han, Clara. 2012. *Life in Debt: Times of Care and Violence in Neoliberal Chile.* Berkeley: University of California Press.

Haugerud, Angelique. 2010. 'Neoliberalism, satirical protest, and the 2004 U.S. presidential campaign.' In *Ethnographies of Neoliberalism*, edited by Carol Greenhouse, 112–127. Philadelphia: University of Pennsylvania Press.

Herzfeld, Michael. 1992. *The Social Production of Indifference: Exploring the Symbolic Roots of Western Bureaucracy.* New York: Berg.

Herzfeld, Michael. 2005. *Cultural Intimacy: Social Poetics in the Nation-State*, second edition. New York: Routledge.

Jalili, Vahid. 2014. *The Islamic Revolution Cultural Front*. Tehran, Iran: The Film Festival of Ammar Press.

Klumbyte, Neringa. 2011. 'Political intimacy: Power, laughter, and coexistence in late Soviet Lithuania.' *East European Politics and Societies* 25 (4): 658–677.

Knight, Daniel M. 2015. 'Wit and Greece's economic crisis: Ironic slogans, food, and anti-austerity sentiments.' *American Ethnologist* 42 (2): 230–246.

O'Neill, Kevin Lewis. 2013. 'Left behind: Security, salvation, and the subject of prevention.' *Cultural Anthropology* 28 (2): 204–226.

Rechtman, Richard. 2017. 'From an ethnography of the everyday to writing echoes of suffering.' *Medicine Anthropology Theory* 4 (3), Special Section, On Affliction: 130–142.

Rose, Nikolas and Peter Miller. 1992. 'Political power beyond the state: Problematics of government.' *British Journal of Sociology* 43 (2): 173–205.

Siahooi, Hamid Reza. 2011. 'Analytical approach to urban informal settlements in Bandar Abbas.' *Geographical Quarterly of Environmentally Based Territorial Planning* 12: 15–45.

Singh, Bhrigupati. 2011. 'Agonistic intimacy and moral aspiration in popular Hinduism: A study in the political theology of the neighbor.' *American Ethnologist* 38 (3): 430–450.

Steinmüller, Hans. 2013. *Communities of Complicity: Everyday Ethics in Rural China*. New York: Berghahn.

Index

Printed in the United States
by Baker & Taylor Publisher Services